## IC 9462
INFORMATION CIRCULAR/2002

# Review of Technology Available to the Underground Mining Industry for Control of Diesel Emissions

**Department of Health and Human Services**
Centers for Disease Control and Prevention
National Institute for Occupational Safety and Health

Information Circular 9462

# Review of Technology Available to the Underground Mining Industry for Control of Diesel Emissions

By George H. Schnakenberg, Jr., Ph.D., and Aleksandar D. Bugarski, Ph.D

U.S. DEPARTMENT OF HEALTH AND HUMAN SERVICES
Public Health Service
Centers for Disease Control and Prevention
National Institute for Occupational Safety and Health
Pittsburgh Research Laboratory
Pittsburgh, PA

August 2002

## ORDERING INFORMATION

Copies of National Institute for Occupational Safety and Health (NIOSH)
documents and information
about occupational safety and health are available from

NIOSH–Publications Dissemination
4676 Columbia Parkway
Cincinnati, OH 45226-1998

FAX: 513-533-8573
Telephone: 1-800-35-NIOSH
(1-800-356-4674)
E-mail: pubstaft@cdc.gov
Web site: www.cdc.gov/niosh

*This document is the public domain and may be freely copied or reprinted.*

Disclaimer: Mention of any company or product does not constitute endorsement by NIOSH.

**DHHS (NIOSH) Publication No. 2002-154**

# CONTENTS

Page

| | |
|---|---:|
| Abstract | 1 |
| Executive summary | 2 |
| 1 Introduction | 4 |
|    1.1 Health concerns | 4 |
|    1.2 Diesel exhaust composition | 4 |
|    1.3 Occupational exposure | 6 |
|    1.4 Selected regulatory limits | 6 |
|    1.5 Diesel particle concerns | 8 |
|    1.6 Document purpose | 9 |
|    1.7 Scope | 9 |
| 2 Control technologies | 9 |
|    2.1 Maintenance | 10 |
|    2.2 Engine design and selection | 11 |
|       2.2.1 Lower DPM emission engines | 11 |
|       2.2.2 Engine derating | 13 |
|    2.3 Fuels | 14 |
|       2.3.1 Commercial fuel and the effects of sulfur | 15 |
|       2.3.2 Alternative fuels | 16 |
|       2.3.3 Fuel additives | 20 |
|    2.4 Aftertreatment technologies | 23 |
|       2.4.1 Diesel oxidation catalytic converters (DOCCs) | 23 |
|       2.4.2 Diesel particulate filters (DPFs) | 26 |
|       2.4.3 DOCC and DPF combinations | 33 |
|       2.4.4 Disposable diesel exhaust filter (DDEF) | 34 |
| 3 Conclusions and recommendations | 35 |
| References | 42 |
| Appendix A.—Low-PI MSHA-approved engines | 48 |

## ILLUSTRATIONS

| | |
|---|---:|
| 1. Monolithic catalyst substrates | 24 |
| 2. Catalyst washcoat | 24 |
| 3. Example of a ceramic monolith filter element | 26 |
| 4. Gas flow in a monolith, wall-flow filter | 26 |
| 5. Knitted microfiber filter (deep-bed filter) | 27 |
| 6. Example of a particulate filter system using fiber media | 27 |

## TABLES

| | |
|---|---:|
| 1. Typical occupational DPM exposure levels | 6 |
| 2. Exposure limits for DPM | 7 |
| 3. Isuzu-Deutz engine comparison | 13 |
| 4. Derated Isuzu engine comparison | 14 |
| 5. Minimum temperature for continuous regeneration | 22 |
| 6. Performance of the available control technologies | 37 |
| A-1. List of MSHA-approved engines (as of December 2001) | 48 |

# ABBREVIATIONS USED IN THIS REPORT

| | |
|---|---|
| ACGIH | American Conference of Governmental Industrial Hygienists |
| CARB | California Air Resources Board |
| CDPF | catalyzed diesel particulate filter |
| CFR | Code of Federal Regulations |
| CONCAWE | Oil Companies' European Organisation for Environment, Health and Safety |
| CRT | Continuously Regenerating Trap |
| DDEF | disposable diesel exhaust filter |
| DECSE | Diesel Emission Control–Sulfur Effects Program |
| DEEP | Diesel Emissions Evaluation Program |
| DETR | Department of the Environment, Transport and the Regions (U.K.) |
| DOC | diesel oxidation catalyst |
| DOCC | diesel oxidation catalytic converter |
| DPF | diesel particulate filter |
| DPM | diesel particulate matter |
| DST | Dry System Technology |
| EC | elemental carbon |
| EGR | exhaust gas recirculation |
| ELPI | electrical low-pressure impactor |
| EMA | Engine Manufacturers' Association |
| EPA | Environmental Protection Agency |
| FBC | fuel-borne catalyst |
| F-T | Fisher-Tropsch |
| FTP | Federal test procedure |
| HC | hydrocarbon(s) |
| ISO | International Organization for Standardization |
| LS | low-sulfur |
| MECA | Manufacturers of Emission Controls Association |
| MOUDI | Micro-orifice uniform-deposit impactor |
| MSHA | Mine Safety and Health Administration |
| NIOSH | National Institute for Occupational Safety and Health |
| OC | organic carbon |
| OEM | original equipment manufacturer |
| OICA | Organisation Internationale des Constructeurs d'Automobiles |
| PAH | polynuclear aromatic hydrocarbons |
| PAS | photoelectric aerosol sensor |
| PDEAS | personal diesel exhaust aerosol sampler |
| PI | (MSHA) particulate index |
| PM | particulate matter |
| PRL | Pittsburgh Research Laboratory (of NIOSH) |
| RCD | respirable combustible dust |
| SAE | Society of Automotive Engineers |
| SMMT | Society of Motor Manufacturers and Traders Ltd. (U.K.) |

| | |
|---|---|
| SMPS | scanning mobility particle sizer |
| SOF | soluble organic fraction (of DPM) |
| SOL | solid fraction (of DPM) |
| SSPD | Sasol slurry phase distillate |
| TC | total carbon |
| TLV® | threshold limit value (ACGIH) |
| TPM | total particulate matter |
| ULSF | ultralow sulfur fuel (<50 ppm of sulfur by weight) |
| VERT | Verminderung der Emissionen von Realmaschinen im Tunnelbau |

## GLOSSARY

| | |
|---|---|
| category A | MSHA designation for engines certified for use inby |
| category B | MSHA designation for engines certified for use in outby areas or metal and nonmetal mines |
| inby | refers to areas of a coal mine that are ventilated by air that are downstream from the point of coal extraction and thus may contain methane; electrical and diesel equipment in this area must be "permissible" and certified as such by MSHA |
| outby | refers to areas of a coal mine that are ventilated by air that has not yet passed the point of coal extraction; either "permissible" or "nonpermissible" diesel equipment can be used here |

## UNIT OF MEASURE ABBREVIATIONS USED IN THIS REPORT

| | |
|---|---|
| bpd | barrel(s) per day |
| cfm | cubic foot (feet) per minute |
| gal | gallon(s) |
| g/bhp-hr | gram(s) per brake horsepower-hour |
| g/hr | gram(s) per hour |
| hp | horsepower |
| hr | hour(s) |
| $\mu g/m^3$ | microgram(s) per cubic meter (mass concentration) |
| $\mu m$ | micrometer(s) ($10^{-6}$ meter) |
| mbar | millibar(s) (which equals 0.1 kilopascal or about 0.7500617 mm mercury, 0.01450284 pounds per square inch pressure) |
| $mg/m^3$ | milligram(s) per cubic meter (mass concentration) = 1,000 $\mu g/m^3$ |
| min | minute(s) |
| nm | nanometer(s) ($10^{-9}$ meter) |
| ppm | part(s) per million |
| rpm | revolution(s) per minutes |
| vol % | volume percent |

# REVIEW OF TECHNOLOGY AVAILABLE TO THE UNDERGROUND MINING INDUSTRY FOR CONTROL OF DIESEL EMISSIONS

By George H. Schnakenberg, Jr., Ph.D.,[1] and Aleksandar D. Bugarski, Ph.D.[2]

## ABSTRACT

This report reviews the performance and applicability of technology for the control of emissions from diesel-powered equipment used in underground coal and metal/nonmetal mines. The methods discussed include Mine Safety and Health Administration-approved low-emission engines, engine derating, fuels, fuel additives, diesel oxidation catalysts, and diesel particulate filters. The potential of each of these technologies is examined individually and in combination. The performance estimates are derived from the published literature and presented in narrative and tabular form. The purpose of this report is to help the mining industry select the most appropriate method to reduce underground exposures of miners to diesel exhaust in the context of the recently developed diesel regulations. It is important to note that the control technologies discussed in this report have received limited evaluation in underground mines. Additional research is ongoing, and some engineering design changes may need to be implemented before all of these diesel emission control technologies can be safely and successfully used in underground mines.

---

[1]Research physicist.
[2]Senior associate fellow.
Pittsburgh Research Laboratory, National Institute for Occupational Safety and Health, Pittsburgh, PA.

## EXECUTIVE SUMMARY

This report presents the potentials and limitations of control technology for reducing exhaust tailpipe emissions from diesel-powered equipment used in underground mining. It does not discuss ventilation, enclosed cabs, or personal protection (respirators), but only commercially available products that reduce the particulate matter (PM) and harmful gases from the exhaust pipe of diesel-powered equipment. While the current operational conditions for diesel equipment commonly provide adequate ventilation to control the harmful gases, these same conditions result in levels of worker exposure to diesel particulate matter (DPM) that are significantly higher than those of any other occupation and significantly higher than the new U.S. and current European standards. Because of a historical concern over the health effects from long-term exposure to DPM and the recent promulgation of two DPM rules by the Mine Safety and Health Administration (MSHA), the focus has been on solutions that reduce PM rather than the toxic exhaust gases. However, as we show, the existing technology can provide significant reductions to both.

The control technology found to be appropriate for discussion falls into the following six general categories: MSHA-approved engines with low emissions, derated engines, fuels, fuel additives, diesel oxidation catalytic converters (DOCCs), and diesel particulate filters (DPFs).

Using low-emission, MSHA-approved engines is a viable option that can result in a significant reduction of PM emissions and, in some cases, a reduction of toxic gas emissions. This option can be demonstrated using the Isuzu C240 (QD60), a very popular engine for outby applications in coal mines. The direct substitution of the Isuzu C240 with a Deutz F4L1011 would result in a 64% reduction of emitted DPM based on the MSHA particulate index (PI) for each engine.

It is possible, and sometimes practiced, to limit the maximum fueling rate of a particular engine to less than its rated maximum specifically to reduce emissions. This practice is called derating. Relatively minor reductions in rated power may result in significant reductions in PM emissions. For example, by reducing the maximum deliverable horsepower of the Isuzu C240 from 56 to 52, DPM emissions are reduced by 62% (from 9.35 to 4.25 g/hr).

The use of commercially available alternative fuels—e.g., biodiesel and synthetic diesel—and water-fuel emulsions can result in lower gaseous and PM emissions. The sulfur content of the fuel becomes a concern when considering the application of oxidation catalysts to the point that the use of ultralow sulfur fuel (ULSF) is highly recommended. Fuel additives, developed as combustion enhancers and smoke reducers, can contain metal compounds. These additives create additional emissions of metallic ash with possible adverse health effects and thus are not recommended. However, some additives, notably fuel-borne catalysts (FBCs), are specifically formulated to lower the exhaust temperatures necessary to burn off the soot collected by particle filters. When used with a particle filter, the FBCs pose no additional health hazard, since the filter effectively prevents emissions of the catalyst metals.

Diesel oxidation catalysts can significantly reduce both carbon monoxide (CO) and hydrocarbons (HC) (the source of the "diesel" odor). They also reduce the organic fraction of PM. However, oxidation catalysts greatly enhance sulfate formation from the sulfur present in the fuel and thereby add another toxic component to the exhaust emissions. Diesel exhaust also contains nitrogen oxides, $NO_x$ ($NO + NO_2$); catalysts tend to convert some of the nitric oxide (NO) to a more toxic form, nitrogen dioxide ($NO_2$), which can be of concern depending on the amount of NO converted and the ability of the prevailing environment to dilute or adsorb (on mine surfaces) the $NO_2$.

DPFs are extremely effective at removing PM (primarily carbon soot and adsorbed HC) from engine exhausts. In addition, the DPFs can be catalyzed to provide good reductions of CO and HC. However, the DPM that is collected by the particle filter needs to be removed. The preferred approach is to use the high exhaust temperatures to burn off the collected soot during engine operation—that is, to cause spontaneous or autoregeneration. As a general rule, the engine should operate at medium to high loads for at least 20% to 25% of the time to ensure sufficient temperature for regeneration. Catalysts are used to lower regeneration temperatures, and their use ensures regeneration under a wide range of applications. Alternatively, filters with integral heaters that are connected to an air and electrical supply managed by an off-board system have been shown to be viable for application in which conditions do not favor autoregeneration. For low-horsepower, light-duty applications, a system that performs an on-board regeneration in 10 min is available. As a final alternative, a DPM-loaded filter can be exchanged for a clean filter that was regenerated in a simple electric kiln designed for the purpose.

Combinations of these technologies provide better reductions of both the gaseous and particulate exhaust components. In particular, the systems combining catalytic oxidation and filtration offer the highest reductions of DPM and gases, except for $NO_2$. This combination produces the best results when used with ULSF.

NIOSH considers the combination of a low-PM-emitting engine (possibly derated), a form of oxidation catalyst, and a DPF, when operated within an effective maintenance program, to be the *best available technology* for reducing hazardous diesel emissions for applications in underground mines. It would be necessary to add the measurement of tailpipe DPM (by methods yet to be selected) to the accepted practice of monthly tailpipe gas checks to ensure that the DPF is performing as expected.

The area of diesel emissions control technology is changing rapidly. The information presented is current as of December 2001. The best source for monitoring this changing field is the World Wide Web. The Web sites to consult include www.cdc.gov/niosh/mining, www.msha.gov/S&HINFO/DIESEL.HTM, and www.deep.org for mining specific information; and www.DieselNet.com and arbis.arb.ca.gov/toxics/diesel/ss/summary_2.htm for developments in particulate emissions control for automotive, highway, and off-highway equipment. The authors of this document may also be contacted.

# 1 Introduction

## 1.1 Health Concerns

Continual expansion of the use of diesel engines in the mining industry and the uncertainties associated with the long-term effects of exhaust emissions on miners' health have recently focused attention on risk assessments of diesel engine exhaust. The issue of DPM has raised a host of health and environmental concerns and has gained considerable attention of researchers and various regulatory agencies.

The concerns about the health effects of DPM exposure have resulted in the issuing of regulations by MSHA governing the emission rates for diesel equipment used in U.S. coal mines [66 Fed. Reg.[3] 5526 (2001)] and environmental compliance levels for DPM in U.S. metal and nonmetal mines [66 Fed. Reg. 5706 (2001)]. Other countries have also adopted regulations that limit environmental levels of DPM [DieselNet 2001]. It is clear from the documentation supporting the MSHA regulations that miners (and other workers in confined spaces) are the most highly exposed of any occupation.

## 1.2 Diesel Exhaust Composition

The complexity of the chemical and physical composition of diesel exhaust emissions makes the assessment of health risk from exposure to diesel engine exhaust a daunting task. Although the major portion of diesel exhaust contains nitrogen, oxygen, water, and the asphyxiant carbon dioxide ($CO_2$), it also contains recognized noxious, toxic, and potentially harmful substances: particulate matter (PM) or soot; organic compounds such as lubricating oil and unburned or partially burned HC, which are primarily the source of the unpleasant odor associated with burning diesel fuel; oxides of nitrogen (NO and $NO_2$, collectively known as $NO_x$); CO; and sulfur oxides. Inorganic constituents of diesel exhaust, such as metals, acids, and salts, are also among the chemical constituents hypothesized to be toxic.

MSHA regulations at 30 CFR 7 define DPM as "any material collected on a specified filter medium after diluting exhaust gases with clean, filtered air at a temperature of $\leq 125$ °F (52 °C), as measured at a point immediately upstream of the primary filter. This material is primarily carbon, condensed HC, sulfates, and associated water." This pragmatic definition provides engine testing facilities with a tool for assessing PM emissions simply by measuring the weight (mass) gain of the filter, i.e., by gravimetric analysis. In practice, physical and chemical processes governing the formation and transformation of the DPM before and after collection make accurate interpretation of DPM emissions and their health impact a complex issue. For example, despite the varying chemical composition of the DPM, the masses obtained are all considered "equivalent" whether they are organic condensates, solid carbon particles, sulfates plus water, or metallic ash from lubrication oils. Using the gravimetric determination of DPM may be useful for comparing engine emissions under nearly identical set of operating conditions, fuels, and measurement methods. However, it is flawed when used to compare the performance of some control technologies and equally inadequate in providing an assessment of health impact. Adding to the complexity is that the prescribed MSHA testing to determine PM emissions is performed over a set of eight steady-state engine operating conditions, which does not therefore include acceleration transients (a major source of PM) and may

---

[3]*Federal Register.* See Fed. Reg. in references.

not be representative of the actual engine operating conditions. According to Kittelson [1998], typical particle composition of a heavy-duty diesel engine tested under a heavy-duty transient cycle breaks down as follows: carbon, 41%; unburned fuel, 7%; unburned oil, 25%; sulfate and water, 14%; and ash and other components, 13%.

The structure and chemical composition of DPM is a function of numerous parameters; the major ones are fuel composition, engine design, engine operating conditions, and the exhaust aftertreatment process. The composition of exhaust particles also depends on where and how they are collected or measured. In addition, the composition of DPM might be significantly altered (and thus different from that measured in the laboratory) when the exhaust is released in the mine setting.

The diameter of diesel particles ranges from 5 nm (1 nm = $1 \times 10^{-9}$ m) to 1 $\mu$m (1 $\mu$m = $1 \times 10^{-6}$ m), depending on engine design and operating conditions. Two distinct size modes characterize PM distribution: the agglomeration mode and the nucleation mode. Particles in the agglomeration mode (50 nm to 1 $\mu$m) contribute to the majority of the DPM mass. The chemical composition of agglomeration-mode particles are mainly a carbonaceous core and adsorbed organic compounds. The nucleation mode contains the majority of the particle number, but does not contribute significantly to the total particulate matter (TPM) mass. Particles in the nucleation mode have been found to be composed mostly of volatile or semivolatile organic compounds, sulfur compounds, and trace elements.

Hydrocarbons, one of the major organic pollutants in diesel exhaust, are emitted as gaseous and PM-bound compounds. The phase (gas, condensed liquid, or solid) of HC in diesel exhaust depends on their molecular weight, temperature, and concentration and is expressed as a partition coefficient, which is the ratio of the mass of compound in the particulate phase to the mass of compound in the vapor phase. Higher molecular weight and some intermediate molecular weight compounds known as soluble organic fraction (SOF) are adsorbed on the PM. Common particulate-bound compounds are linear- and branched-chain HC with 14 to 35 carbon atoms; polynuclear aromatic hydrocarbons (PAHs); alkylated benzenes; nitro-PAH; and a variety of polar, oxygenated PAH derivatives. These compounds are of particular concern because several of them have been associated with carcinogenicity and mutagenicity, as reported in various laboratory studies. Some of the vapor phase compounds that could potentially affect human health include formaldehyde, methanol, acrolein, benzene, 1,3-butadiene, and low-molecular-weight PAHs and their oxygenated and nitrated derivatives.

Several methods are in use for determining occupational exposures to the components of diesel exhaust. Direct-reading instrument and chemical methods (stain tubes) are used to determine the gas concentrations of $CO_2$, CO, NO, and $NO_2$. Field determination of the organic vapor fraction is not normally performed. The determination of workplace concentrations of exhaust PM is a more complex undertaking. There is no consensus on exactly what is to be measured (total or elemental carbon, combustible carbon) or the analytical method (thermo-optical, coulometric, gravimetric) for determining workplace DPM concentrations. A synopsis of these methods is presented in the MSHA "Diesel Toolbox" [MSHA 1997]. Gravimetric-based methods that determine the amount of respirable combustible dust (RCD) [Maskery 1978; Gangal et al. 1990; Gangal and Dainty 1993] or that determine the amount of submicron particulate collected on a filter are inadequate at moderately low workplace concentrations because so little mass is collected that weighing error becomes significant. Methods based on determining the amount of elemental carbon (EC) present, such as the coulometric method [ZH 1/120.44 (1995)] adopted in Europe and the NIOSH Method 5040 [NIOSH 1999], are much more sensitive to DPM. These methods exploit the established unique and reasonably strong correlations between EC and DPM [NIOSH 1999]. In coal mines, because of the presence of both organic and EC in the coal mine dust, a submicron impactor [Cantrell et al. 1993] can be used to separate the larger dust particles from the diesel combustion particles for EC analysis. A direct-reading instrument, such as the Ecochem PAS 2000 photoelectric aerosol sensor (PAS), is available and can be used for noncompliance measurement of workplace or tailpipe (with dilution) carbon particle concentration. The state-of-the-art aerosol instrumentation used for measurement of mass (MOUDI, ELPI, etc.) and number (SMPS) concentrations is too costly, complex, and cumbersome for routine monitoring of PM underground.

### 1.3 Occupational Exposure

Table 1 shows the occupational exposures to DPM of miners and of those in other occupations.

**Table 1. — Typical occupational DPM exposure levels**

| Occupational group | Exposure level, $mg/m^3$ (1 $mg/m^3$ = 1,000 $\mu g/m^3$) |
|---|---|
| Underground miners, coal, no aftertreatment[1] | 0.9 - 2.1 |
| Underground miners, coal, disposable diesel exhaust filter[1] | 0.1 - 0.2 |
| Underground miners, coal, wire mesh filter[1] | 1.2 |
| Underground miners, metal/nonmetal, no aftertreatment[1] | 0.3 - 1.6 |
| Surface miners[1] | <0.2 |
| Urban fire station[2] | 0.1 - 0.48 |
| Forklift operators, dock workers, railroad workers[2] | 0.02 - 0.10 |
| Truck drivers[2] | 0.004 - 0.006 |

[1]Haney et al. [1997].
[2]DieselNet [1999b].

### 1.4 Selected Regulatory Limits

Table 2 provides some regulatory limits for occupational exposures to DPM.

Table 2.—Exposure limits for DPM [DieselNet (2001)]

| Country or organization | Value, mg/m$^3$ | DPM Measurand |
|---|---|---|
| **Current Limits:** | | |
| U.S.: MSHA metal/nonmetal underground mines [66 Fed. Reg. 5706 (2001)] | July 19, 2002: 0.4<br>January 19, 2006: 0.16 | Total carbon (EC + OC) as determined by NIOSH Method 5040 |
| U.S.: MSHA underground coal mines [66 Fed. Reg. 5526 (2001)] | Emissions rates set for various classes of equipment, e.g., heavy duty equipment: 2.5 g/hr | Total DPM measured in accordance with ISO 8178 procedures [30 CFR[4] 7 (1996)] |
| Germany: General occupational environment | 0.1 | EC, coulometric |
| Germany: Underground metal and nonmetal mining and construction sites | 0.3 | EC, coulometric |
| Canada: Underground, metal and nonmetal mining | 1.5 | RCD |
| Switzerland [Majewski 1999] | 0.1 | EC, coulometric |
| **Proposed Limits:** | | |
| ACGIH [1995] | 0.15 | Particles <1 $\mu$m in size |
| ACGIH (1998) | 0.05 | Total carbon in particles <1 $\mu$m in size |
| ACGIH (2001) | 0.02 (EC = 40% of DPM) | EC particles <1 $\mu$m in size |

"In its 2001 Notice of Intended Changes, the ACGIH proposed a TLV of 0.02 mg/m$^3$ for diesel exhaust particulates measured as elemental carbon (EC), with proposed carcinogenicity classification A2 - "Suspected Human Carcinogen." This EC-based TLV is practically equivalent to the previously proposed TLV of 0.05 mg/m$^3$, presumably as total diesel particulate matter (EC fraction typically constitutes about 40% of the total diesel particulate mass)." (Source: www.DieselNet.com.)

In January 2001, MSHA promulgated two new rules regulating the exposure of underground miners to DPM. The metal rule [66 Fed. Reg. 5706 and 35518 (2001)] requires underground metal and nonmetal mine operators to comply by July 19, 2002, with the interim DPM concentration of 400 $\mu$g/m$^3$ measured as total carbon using NIOSH Analytical Method 5040 [NIOSH 1999]. On January 19, 2006, the compliance level DPM concentration limit will be 160 $\mu$g/m$^3$ measured as total carbon.

The underground coal rule [66 Fed. Reg. 5526 and 27864 (2001)] controls the exposure of the miners by limiting the emission rate from newly introduced and existing diesel-powered

---

[4]*Code of Federal Regulations.* See CFR in references.

equipment. MSHA found that due to the absence of an accurate method for sampling DPM in underground coal mines, a performance rule, similar to that promulgated for metal and nonmetal mines, is not feasible. Additionally, in-place coal regulations [30 CFR 75.325(f) and (g) (1996)] specify ventilation air quantities for each piece of equipment, allowing reasonably precise estimation of the resulting DPM concentration from the engine emissions. Engine emission rates are calculated using the emission rate determined through MSHA engine certification [30 CFR 7 (1996)] and the reduction provided by the application of emission control technology, namely, particle filters. On its Web site, MSHA provides a list of filters and their accepted filtration performance for this purpose [MSHA 2001a]. Should alternative emission controls be used, the emission rates of the engine and control system would have to be determined using the MSHA engine certification procedures [30 CFR 7 (1996)]. Only systems verified as meeting the 2.5 g/hr (permissible or heavy-duty) or 5 g/hr limits (outby light-duty) would be accepted.

The coal rule contains a complex timetable of equipment type and emission rates. Suffice it to state that all heavy-duty equipment must eventually meet a DPM emission rate of $\leq 2.5$ g/hr. Newly introduced light-duty equipment must either use an engine approved by the Environmental Protection Agency (EPA) or emit <5 g/hr of DPM. Existing light-duty equipment is exempt from the rule. The MSHA Web site lists MSHA-approved engines and their emission rates [MSHA 2001b].

Both rules leave the choice of controls up to the mine operators. The metal and nonmetal rule allows a wider choice of control technology, since it uses an environmental standard to measure compliance. The choices of control technology for coal are much more limited, if for no other reason than economic: any system that does not use a device with an MSHA-accepted PM reduction factor must undergo expensive testing. Also, for coal mines, only a limited number of engines and associated power packages are available that are suitable for use in gassy areas of the mines.

**1.5  Diesel Particle Concerns**

Current PM emissions legislation is based on ambient mass concentrations, $mg/m^3$. None of these regulations contain a reference to either the size or the number concentration of the particles. Further, the prescribed gravimetric analysis of PM is nonspecific with respect to chemical composition and aerosol properties and, thus, delivers no toxicologically relevant information. Additionally, known DPM size distributions indicate the presence of very fine particles that, when inhaled, are eventually trapped in the slowly cleared alveolar regions of the human lungs. The health implications of these ultrafine particles is currently unknown and is the subject of much speculation and research. The MSHA DPM rule for metal and nonmetal mines [66 Fed. Reg. 5706 and 35518 (2001)] uses total carbon as measured by NIOSH Method 5040 [NIOSH 1999] as a compliance measurement. This method, however, accounts for neither particle size nor inorganic continuants such as sulfates and transitional metals.

The effects of inorganic constituents of DPM on mortality and morbidity have been the subject of numerous epidemiological and toxicological studies [Mauderly et al. 1995] and should not be underestimated. The transition metals can cause the production of hydroxyl radicals, which are considered to be toxic products. Residual lubrication oil ash is also toxic to cells and lungs. Finally, a wide range of inorganic and organic sulfur and nitrogen compounds have irritating, cytotoxic, and mutagenic properties. Animal studies [Schlesinger 1995] indicate that nitrates, sulfates, and sulfuric acid particles impair pulmonary functions such as mucociliary clearance and airway resistance.

## 1.6 Document Purpose

The purpose of this document is to provide an overview and review of practical, available technology that can be used to reduce gaseous and particle emissions from new or older diesel-powered equipment in underground mines. Upon reading this document, it is hoped that the decision-makers in labor, industry, and government will be able to learn which control alternatives are available and proven, to what extent each is able to reduce toxic gases or harmful PM from diesel exhaust, the caveats and conditions of use for each, the effects of combining technologies, and the estimated costs. Armed with this knowledge, it is additionally hoped that the same entities will be able to recognize the most effective technologically feasible controls for reducing miners' exposure to diesel emissions in underground mines.

## 1.7 Scope

This document presents the performance and limitations of control technology designed to reduce diesel exhaust emissions from the tailpipe. It does not discuss ventilation, enclosed cabs, personal protection (respirators), or measurement technology, but only proven commercially available technology that reduces the PM and toxic gases from the exhaust of diesel-powered equipment.

The technology discussed is applicable to most, if not all, diesel equipment used in underground coal or metal/nonmetal mines. Additional design and engineering efforts must be made to adapt some of the technology for use in areas (inby) of coal mines that require precautions against methane ignition or hot surface temperatures.

This document does not address maintenance or proper application of diesel engines. It is expected that the industry has or will shortly institute across-the-board maintenance procedures that, at a minimum, follow MSHA guidelines and the engine or vehicle manufacturer's prescribed maintenance procedures. Additionally, engines, when operated at altitudes >1,000 ft, need to be properly derated, and then the torque converter of the vehicle needs to be matched to the derated engine to avoid excessive emissions from lugging down the engine and to allow the engine to attain the optimum engine speed for maximum power transfer to the drive train.

The control technology found to be appropriate for discussion falls into the following general categories: low-emission engines, derated engines, fuels, fuel additives, DOCCs, and DPFs. Combinations of these technologies are possible and are also discussed.

## 2 Control Technologies

Mine ventilation has traditionally been the primary means for controlling workplace concentrations of diesel emissions in underground mines. With the continual increase in the number of diesel equipment units deployed in underground mines and the rising concerns about adverse health effects of diesel emissions, increasing the ventilation rate as the sole means to control exposures becomes an inadequate and expensive approach. Concerns with engine efficiency and the general environmental impact of diesels on urban air quality have driven research to reduce emissions at the source, resulting in much lower-emitting (particularly NO$_x$ and PM) diesel engines and practical aftertreatment and other emission reduction technologies. Although some of the aftertreatment technology was originally developed for use in underground mines and tunnels, the present drive for research results from the need to address on-highway diesels. (The latest information on worldwide regulations can be found on at www.DieselNet.com/standards.html; the current EPA rules can be found at www.epa.gov/otaq/diesel.htm.) The underground mining industry can benefit greatly from the advances in technology and the economies of scale of the on-highway truck and bus market. This document provides a brief review of the most effective technologies developed for curtailment of diesel emissions, including the toxic gases and PM. These technologies include engine design, engine derating, fuel formulations, fuel additives, DOCCs, and DPFs.

Evaluation of the performance of emission control technology in underground mine settings is difficult because of the intrinsically transient operation of machinery with the pronounced variations in numerous parameters, such as daily workload and ventilation rates. Therefore, only a very limited amount of accurate data exist on the emission reduction performance of the technologies in underground mine conditions. Nonetheless, the examination herein of the body of literature representing laboratory tests and the knowledge of the scientific and engineering principles of the technology provide an adequate foundation for estimating the effects of the application of this technology to reduce diesel emissions in the underground mine operations. This analysis also provides sufficient data to support decisions on the technical feasability of control technology alternatives. However, the limited field performance data available point to the need to perform more field evaluations on appropriately selected technology.

## 2.1 Maintenance

Over the relatively short history of the use of diesels in underground mining, the need for good maintenance has always been recognized. This recognition has not always generated the adoption and disciplined implementation of the best maintenance practices being applied to every diesel engine or vehicle in operation in underground mines. Nevertheless, it is extremely important to realize that the very first step on the path to reducing worker exposures is to implement an effective diesel vehicle/engine maintenance protocol and apply it to *every* diesel unit that operates underground. The early work in this area was performed by the U.S. Bureau of Mines [Waytulonis 1987]. The University of Minnesota's Center for Diesel Research [Spears 1997] developed procedures for using tailpipe gas measurements as a diagnostic for engine maintenance. A comprehensive study on the relationship between diesel engine maintenance and tailpipe emissions was recently completed by McGinn [2000] under a research effort by the Diesel Emissions Evaluation Program (DEEP). McGinn has developed a maintenance auditing procedure [McGinn et al. 2000] and guidelines [McGinn 1999], which were recently implemented in a hard-rock mine with demonstrable results. The guidelines and training of the mine personnel involved participation by mine management, machine operators, mechanics, and most importantly the engine and vehicle manufacturers' service (not sales) representatives. Dramatic reductions in exhaust PM and CO emissions were observed in some cases where good maintenance practice was applied. These documents can be found at www.deep.org/research.html.

There are several important reasons to provide the best possible engine maintenance when considering or implementing control technology. The first reason is that the lowest emissions resulting from the application of any control technology are obtained when starting with the lowest possible engine-out emissions. The second reason is that ventilation requirements and PM emission rates determined through MSHA's engine certification process were obtained using a properly tuned, well-maintained (new) engine. It is important that the engines in the field have emission characteristics no worse than those of the certified engine so that calculations that use the MSHA ventilation and PM emission rates to ensure safe levels of toxic gases, to estimate workplace diesel particulate levels, or to compare particulate emission rates among engines are valid. In addition, excessive emissions from poorly maintained engines may jeopardize performance of aftertreatment technologies. For example, excessive emissions of the ash caused by burning crankcase oil might result in clogging and premature failure of the DPF.

In sum, although maintenance is not strictly an "add-on" hardware control technology, the very first step in applying control technology to reduce workplace exposures to diesel exhaust is to implement an effective maintenance program and closely monitor its effectiveness.

## 2.2 Engine Design and Selection

Over the last decade or so, major improvements have been made in engine design that have resulted in substantially lower emission rates of DPM. Additionally, lower engine emissions can be obtained by limiting the maximum fueling rate to an engine, resulting in a lower power output but substantial reduction in DPM emissions and some fuel savings.

### 2.2.1 Lower DPM Emission Engines

The major efforts in reducing PM emissions from diesel engines have been directed toward optimizing the combustion and fuel injection systems and minimizing the lube oil consumption of the engine. The successful engineering techniques for these purposes are high compression, air intercooling of turbocharged engines, center positioning of the injector nozzle, increased number of the nozzle perforations, very high injection pressure, suppression of the air swirl, and a shallow piston bowl. These techniques have resulted in significant reductions of PM mass emitted from the engine. Mayer [1997] found that modern engines emit 10% of the total particulate mass emitted by engines designed 15 to 20 years ago. Unfortunately, these new low-emission engines were found to emit more ultrafine particles at all load points than the older engine of the same family [Bagley et al. 1993; Baumgard and Johnson 1996; Mayer 1997]. Mayer et al. [1999] concluded that engine designers presently do not have a strategy for effectively curtailing the emission of these nanoparticles. It is yet to be confirmed that this increase in nanoparticles, which was observed under laboratory test conditions, actually manifests itself in actual mine settings. Nevertheless, tests of DPFs using several different filter media confirm that filters are very effective in "trapping" the nanoparticles [Mayer et al. 1999; Czerwinski et al. 1998].

Although a variety of engineering technologies have resulted in engines with greatly reduced emissions, mine operators are presently limited in their choice of diesel engines. After November 24, 1999, only engines listed by MSHA [MSHA 2001b] as approved can be used in coal mines [30 CFR 75.1907 (1996)]. After March 20, 2001, all diesel engines introduced into metal and nonmetal mines must be MSHA- or EPA-approved [30 CFR 57.5067 (2001)]. Furthermore, they must be set to operate at the conditions under which the approval was granted. Thus, with the exception of derating for altitude, there is little or no room for altering engines, let alone changing the design. Furthermore, the MSHA requirement that all engines in use in coal mines must have current approval may be the reason that many engines of older design appear along with the newer engines on the MSHA list.

The engine selection limitation is exacerbated for the permissible areas in coal mines. Only six engines are approved, and these are of much older design. These engines have very high PM emission rates. In some cases, a reduction of 95% in PM is needed to bring the emission rate down to 2.5 g/hr required by the coal rule. As this review will reveal, a 95% reduction in total DPM challenges the capabilities of the contemporary control technology. The limited selection has also resulted in engine applications that underutilize the engine, wasting fuel, and unnecessarily increasing DPM concentrations in the workplace. On the other hand, some applications require more horsepower than these engines can provide, especially when they are derated for altitude. Clearly, there is a need for cleaner permissible engines. Because of the significant expense of certifying an engine for permissible applications and the minuscule market size, there is little incentive for engine manufacturers to provide new permissible engines to the coal industry.

In its 1996 revision to diesel engine approval procedures [30 CFR 7 (1996)], MSHA recognized the concerns over diesel particulate emissions and recognized that newer engines can emit DPM at a much lower rate. Therefore, MSHA provided a means, the PI, for conveying this lower PM emission to the mining industry. The PI is the amount of air needed to dilute the engine-produced DPM to 1 $mg/m^3$. It is calculated from a weighted-average PM obtained over the ISO 8178 C1, eight-mode, steady-state test described in 30 CFR 7. A lower PI characterizes an engine with a lower PM emission. Unfortunately, but unavoidable, the MSHA test procedures do not account for the fact that a significant portion of real-life PM emissions occurs during engine transients (accelerations) or for the significant variation in duty cycles across applications.

The PI and the ventilation rate are the keys for selecting a low-emission engine for underground mine applications. With the understanding that a given application requires an engine with a certain power, rated speed, and physical size, one can examine the MSHA approval list sorted by horsepower (see the appendix to this report) and pick an engine with the lowest PI that closely matches the engine power and rated speed required for the application. Next, one can check the ventilation rate required by that engine and determine whether that rate is acceptable. (Some of the low-PI engines require exceptionally high gaseous ventilation rates that may pose limitations on their use if ventilation is critical). Mine operators should consider these factors when purchasing new equipment or a replacement engine. Likewise, equipment suppliers should try to design their equipment for low-emission engines and offer it as an option in their equipment lines.

A close look at the tabulated nameplate ventilation rates and particulate indices for the category B (nonpermissible or outby) engines approved by MSHA reveals significant differences in the emissions, especially in engines of low horsepower. Table 3 shows the differences between the Isuzu C240(QD60) engine and a comparable Deutz F4L1011 engine.

Table 3.—Isuzu-Deutz engine comparison

| Engine | Rating hp @ rpm | MSHA name plate ventilation rate, cfm | MSHA PI, cfm | MSHA approval | Approximate cost |
|---|---|---|---|---|---|
| Isuzu C240 (QD60) | 56.0 @ 3,000 | 2,500 | 5,500 | 7E-B038-0 | $4,000 |
| Deutz F4L1011 | 56.3 @ 3,000 | 3,000 | 2,000 | 7E-B060-0 | $6,000 |

In MSHA approval tests, the Deutz F4L1011 engine emits, on average, approximately 64% less DPM than the Isuzu C240 engine. Assuming that the prevailing ventilation rate is already >3,000 cfm and remains unchanged, switching to the Deutz engine will reduce the contributions to the workplace DPM concentrations from that vehicle by almost two-thirds. If the ventilation rate must be increased from 2,500 to 3,000 cfm to accommodate the Deutz, then by virtue of this increased ventilation rate, the contribution of the Deutz to the workplace DPM concentration is 70% less than that of the Isuzu.

The example involving the Isuzu and Deutz is not unique; numerous low-PI engines over 100 hp are also available for substitution. Table A-1 in the appendix of this report lists MSHA-approved engines. In this table, the engines with exceptionally low PIs have been identified. Significant workplace reductions, primarily in metal and nonmetal mines, can be achieved by replacing any existing engine having a high PI with a current MSHA-approved engine with a low PI.

It is noteworthy that DPM reductions attainable by selecting clean engines or derating engines can be quite substantial. EPA-certified engines emit one-tenth the DPM than those of 10 years ago. There remains a question (which NIOSH hopes to investigate) about the alleged increase in nanoparticle number from low-PI engines [Bagley et al. 1993; Baumgard and Johnson 1996; Mayer 1997]. Health professionals are concerned and uncertain about the effect of nanoparticles on worker health.

Lastly, for optimum results regardless of the choice made, the engine-out (before any aftertreatment if it is used) emissions of DPM and toxic gases must be kept to the minimum by diligent application of proper maintenance.

### 2.2.2 Engine derating

Substantial reductions in PM emission rates can result from lowering of the maximum fueling rate of an engine. It is possible that many of the engines certified by MSHA are certified at or very near their maximum power where PM emissions are quite high. There is no restriction by MSHA on reducing the maximum power delivered by lowering the maximum fuel setting, i.e., derating. It is not unheard of that some mines choose to derate their engines to reduce DPM and CO emissions, reduce tire slippage and wear, and save on fuel costs. The MSHA list provides some insight into the effects of derating an engine on PM emissions.

Table 4 shows the effects of derating on gaseous and DPM emissions from an Isuzu C240MA engine. The DPM emission rate can be reduced by 55% (calculated from the MSHA PI) with only a 7% reduction in power. If each engine is operated at the nameplate ventilation rate, the resulting DPM concentration is reduced by 62%.

**Table 4.—Derated Isuzu engine comparison**

| Engine | Rating hp @ rpm | MSHA name plate ventilation rate, cfm | MSHA PI, cfm | MSHA approval |
|---|---|---|---|---|
| Isuzu C240MA | 56.0 @ 3,000 | 2,500 | 5,500 | 7E-B085-0 |
| Isuzu C240MA | 52.0 @ 3,000 | 3,000 | 2,500 | 7E-B086-0 |

The comprehensive list of MSHA-approved engines provided on pages 5667-5668 of the coal rule [66 Fed. Reg. 5526 (2001)] have the Isuzu QD100-306 engine listed at 66 hp with a PI of 10,000 cfm and at 70 hp with a PI of 50,000 under the same approval number. Therefore, the reduction of 4 hp by limiting the fueling rate results in an 80% reduction in the DPM emission rate for this engine.

These two examples may be anomalies. PM emission rates for certified engines at powers lower than that of the certification testing may be obtained from the engine manufacturer. If the loss of power would not affect the performance of the equipment for a particular application, it is certainly advisable to check with the engine manufacturer on the emission reductions to be gained by derating the engine. If the derating is substantial, it is also advisable to check with the equipment or torque converter manufacturer to determine whether another converter should be used to obtain an optimum power match to the derated engine. The proper procedures are explained by Forbush [2001].

## 2.3 Fuels

In parallel to the development of the cleaner diesel engine technologies, much attention has been given to the defining of future diesel fuel quality requirements. Extensive research in this field [Baranescu 1988; Cowley et al. 1993; Xiaobin et al. 1996] has shown that properties such as sulfur content, fuel density, cetane number, oxygen, and aromatic content are the physical and chemical properties that most significantly influence particulate and gaseous emissions. Research by Den Ouden et al. [1994] showed that the contribution from fuel properties other than sulfur to heavy-duty diesel emissions is comparatively small and can be characterized by a combination of cetane number and density. Fuel sulfur forms both sulfur dioxide (gas) and sulfates (solids at room temperature). A comprehensive summary discussion and table are presented in the "Diesel Fuels" section of the Technology Guide of the DieselNet Web site [DieselNet 1998d].

### 2.3.1 Commercial Fuel and the Effects of Sulfur

The mechanism for sulfate formation is as follows: during fuel combustion, the sulfur oxidizes to produce sulfur dioxide ($SO_2$), a fraction (<5%) of which can be further oxidized to sulfur trioxide, $SO_3$, which combines with water to form a sulfuric acid aerosol [Heywood 1988]. Studies with low-sulfur fuel revealed that the number of relatively large particles (>0.040 $\mu$m) remains unaffected when fuel with low sulfur content is used. In contrast, low sulfur content is found to reduce the concentration of nanoparticles (<0.040 $\mu$m) by several orders of magnitude, revealing that most particles of this size are sulfur-related. Certain aftertreatment technologies such as diesel oxidation catalysts (DOC) and catalyzed diesel particulate filters (CDPF) can exacerbate the conversion of $SO_2$ to $SO_3$ and thus should be used with ULSFs (<50 ppm) to minimize sulfate particulate formation and poisoning of the catalyst.

Sulfur content in current low-sulfur diesel fuel (Federal LS No. 2 diesel fuel, which is the U.S. on-highway truck fuel) is on average 340 ppm (maximum of 500 ppm) by weight and, most likely, will be reduced further in the future. The sulfur content of diesel fuel currently used in California (CARB diesel) averages 120 ppm S. A recent economic study sponsored by the Engine Manufacturers Association (EMA) concluded that the incremental cost to reduce sulfur level in diesel fuel from the current 500 ppm to <50 ppm would be on average about 5 to 7 cents/gal [EMA 1999]. The Manufacturers of Emission Controls Association (MECA) reported that reducing the level of sulfur in diesel fuel would allow the introduction in the United States of several promising control technologies that reduce emissions of $NO_x$ and PM [MECA 1999b]. In December 1999, ARCO announced plans to offer a cleaner burning diesel fuel to help reduce soot emissions from urban municipal fleets in southern California. The ARCO fuel will have a maximum sulfur content of 15 ppm. The price of the ARCO fuel is expected to be approximately 5 to 7 cents/gal more than that of the CARB diesel fuel (120 ppm S). On December 21, 2000, EPA announced that refiners will be required to start producing diesel fuel for highway use with 15 ppm sulfur or less by June 1, 2006, with availability at retail stations by September 1, 2006.

Klein et al. [1998] studied the effects of fuel sulfur content on the PM emissions from diesel passenger cars equipped with oxidation catalysts. The engines were tested at steady-state conditions. Klein et al. found that with low-sulfur fuel and low exhaust temperature, the PM reduction was linked to a shift in particulate size distribution toward smaller sizes. For low-sulfur fuel and high exhaust temperature, the particulate mass emission rate increased with the use of an oxidation catalyst. This trend was attributed to $SO_3$ production and a shift in particulate size distribution toward larger sizes. For high-sulfur fuel and low exhaust temperature, PM emissions had the same trend as that for low-sulfur fuel and low exhaust temperature. For high-sulfur fuel and high exhaust temperature, PM emissions showed the same trend as that for low-sulfur fuel and high exhaust temperature. Significantly, Carder [1999] found that reducing fuel sulfur from 0.3% (3,000 ppm) to 0.04% (400 ppm) resulted in a 22% reduction of DPM mass emission in an MWM D916-6 engine operated over the ISO 8178 cycle.

The Diesel Emissions Control–Sulfur Effects (DECSE) Program, a joint government/industry research effort, evaluated the impact of diesel fuel sulfur level on the emission control systems such as $NO_x$ absorber catalyst, DPF, lean-$NO_x$ catalyst, and DOC [DECSE 2001]. A study [DECSE 1999] on the effects of diesel sulfur level on DOC performed on the Cummins ISM 370 engine showed that, at high exhaust temperatures (518 °C, OICA Mode 2), the *engine-out* PM emissions are largely independent of fuel sulfur level. PM emissions over heavy-duty FTP cycle varied independently of fuel sulfur levels for both engine-out and catalyst-out emissions. At this condition, engine-out sulfate conversion was approximately 2%. A DOC with relatively low precious metal content (low activity) increased the sulfate conversion to 10%. *Catalyst-out* emissions showed a very strong sulfur effect: an increase in fuel sulfur from 3 to 350 ppm resulted in a 400% increase in PM emissions.

A study [DECSE 2000] of the effects of diesel sulfur level on performance of a CDPF and a Continuously Regenerating Trap (CRT) using a Caterpillar 3126 engine showed that the engine-out PM emissions increased approximately 30% when the fuel sulfur level was increased from 3 to 350 ppm. Both DPFs reduced PM emissions by 95% over the OICA cycle with 3-ppm sulfur fuel. However, with 30-ppm sulfur fuel, the PM reduction efficiencies dropped to 74% and 72% for the CDPF and CRT, respectively. With the 150-ppm sulfur fuel, the postfilter PM emissions increased and efficiencies were 0 and 3%; with the 350-ppm sulfur fuel, the PM emissions increase was 122% and 155% for CDPF and CRT, respectively. The effects of fuel sulfur level on gas-phase emissions and fuel consumption were not significant. Analysis of the PM showed that the increases were attributable to the sulfur content. Nearly 40% to 60% of fuel sulfur was converted to sulfate PM, as measured over the 13-mode OICA cycle for both DPFs.

As a part of DETR/SMMT/CONCAWE Particulate Research Programme, Andersson and Wedekind [2001] compared effects of diesel fuel sulfur content on DPM emissions. They compared diesel fuels with sulfur content of 500 ppm, 300 ppm, and 50 ppm with ultralow sulfur (<10 ppm) Swedish Class I diesel fuel. Ultralow sulfur showed a small, but significant reduction in particle mass and number compared to the other fuels tested. Effects of fuel sulfur were found to be greatest within the nucleation mode particles.

In summary, the sulfur content of diesel fuel adversely affects diesel emissions by producing $SO_2$ and sulfates. The use of oxidation catalysts further increases the production of sulfates, which, unfortunately, are not significantly trapped by particle filters, decreasing their effectiveness to reduce DPM mass and creating the potential for nanoparticle formation [Kittelson 1998]. The rate of sulfate production depends on catalyst formulation and exhaust (catalyst) temperature. Catalysts that are highly effective at converting CO and HC are also highly effective at $SO_2$ conversion. It follows, therefore, that a less active catalyst produces fewer sulfates, but with the penalty of less reduction in CO and HC. At low exhaust temperatures (<225 °C), $SO_2$ conversion to sulfates is minimal, but at higher temperatures (between 225 °C and 560 °C, peaking at 450 °C) conversion is substantial [DieselNet 1999a]. *Since exhaust temperature is not a freely controllable parameter in the field, the production of sulfates and thus workplace concentrations vary greatly and are unpredictable.* Additionally, sulfates poison catalysts. It is clear, therefore, that both $SO_2$ and sulfates add to the toxic burden of the exhaust. For these reasons, it is advisable to know the sulfur content of the fuel and crankcase oil used and to strive to use those with the lowest sulfur content in underground mines. Use of lower sulfur fuels and oils results in lower exhaust toxicity from lower sulfate and permits the use of more active catalysts that are highly effective in reducing CO and HC. DOCC suppliers should take into account the sulfur content of the fuel used at their customers' mines when providing DOCCs for their equipment.

### 2.3.2 Alternative fuels

Reformulated and alternative diesel fuels recently received significant attention as a way of controlling emissions and providing energy independence. Fuels such as biodiesel and synthetic diesel fuel obtained through Fisher-Tropsch (F-T) conversion are high-quality alternative fuels. These fuels can be used in neat form or blended with petroleum diesel fuel to make a cleaner diesel fuel. Fuel-water emulsions also promise reductions in the $NO_x$ and PM emissions.

### 2.3.2.1 Fisher-Tropsch

Fisher-Tropsch (F-T) conversion is a gas-to-liquid process used for synthesis of HC from CO and hydrogen. Historically, the process was used for producing synthetic diesel fuel from coal, natural gas, and biomass resources in countries where petroleum fuel stock was in short supply. The process has received attention recently because of its ability to convert natural gas resources to liquid fuels and chemicals. The synthetic diesel fuel produced by this process is of very high quality and has the potential to significantly reduce exhaust emissions. The F-T fuels have a high cetane number (up to 70), a low sulfur content (<10 ppm), and a low aromatic content (<3%). The benefits are most pronounced in reducing PM emissions [Schaberg et al. 1997; Mayer 1997; Norton et al.1999] owing in part to the almost complete absence of sulfur and its accompanying sulfate emissions. Absence of sulfur also enables the use of catalytic oxidation technologies without the concern over catalyst poisoning or PM emission penalties from the catalytic creation of sulfates. McMillian and Gautam [1998] concluded that F-T fuels provide a basis for reduction of $NO_x$ using higher exhaust gas recirculation (EGR) rates. However, the low efficiency of the F-T process currently makes this fuel expensive. Because the lubricity of F-T fuel is significantly lower than that of regular diesel, the addition of a lubricity improver is required.

F-T fuel is not commercially available in the United States, but companies including Shell, Chevron, Exxon, and ARCO are working on developing production. F-T fuel is commercially available in the Republic of South Africa from companies such as Sasol and Mossgas. Syntroleum Corp. currently owns and operates a pilot plant in Tulsa, OK, where it has successfully demonstrated numerous elements and variations of the Syntroleum Process (an F-T process based on a proprietary catalyst developed by Syntroleum) since 1990. ARCO and Syntroleum are presently building a 70-bpd demonstration plant at ARCO's Cherry Point Refinery in Washington State. Syntroleum and Enron are planning to complete an 8,000- to 10,000-bpd specialty chemical plant late in 2001. Syntroleum is expecting to build a commercial-scale gas-to-liquid plant within the next 3 years.

Recently, the performance of synthetic diesel fuels was examined by several researchers. Schaberg et al. [1997] examined diesel exhaust emissions using Sasol slurry phase distillate (SSPD) process fuel. It was found that the SSPD fuels produce significantly lower emissions than the diesel No. 2 and CARB fuels in all four regulated emission categories. When compared to the No. 2 diesel fuel, HC, CO, $NO_x$, and PM emissions were reduced by 49%, 33%, 27%, and 21%, respectively. The exhaust emissions were lower owing to the very high quality of the synthetic diesel fuel used (cetane number >70, aromatic content <1%, sulfur content <10 ppm). The soluble organic fraction (organic carbon (OC)) of the integrated PM was found to be significantly lower when the cetane number was increased, but this benefit was offset by an increase in the insoluble (EC) portion of the TPM. Schaberg et al. [1997] also found a linear relationship between fuel sulfur and the sulfate portion of total particulate emissions. The influence of F-T fuel on the particle size of the PM was not examined.

Mayer [1997] compared standard Swiss low-sulfur diesel fuel with a chemically pure paraffin fraction made by DEA-Mineralöl AG, Hamburg, Germany (negligible content of sulfur, nitrogen, and aromatics). Emission improvements with synthetic fuel were reported as "disappointingly low." Mayer concluded that the reformulation of diesel fuel cannot efficiently curtail the emission of ultrafine particles.

More recently, Norton et al. [1999] examined regulated emissions from older model transit buses operated on an F-T fuel (produced by Mossgas of the Republic of South Africa) using West Virginia University's transportable chassis dynamometer. Three buses without and three with catalytic converters were tested. Compared to their emissions when operating on No. 2 diesel fuel, buses without catalytic converters emitted 20% lower PM; buses with catalytic converters emitted 31% lower PM when operating on neat Mossgas fuel.

Bugarski [1999] tested an Isuzu C240 engine with diesel No. 2 and neat F-T fuel. It was found that when using the synthetic diesel instead of diesel No. 2 the *mass* of PM emitted decreased. However, the *number* of ultrafine particles, i.e., those particles that are thought to be deposited in the alveolar region of the lungs, unfortunately, increased.

### 2.3.2.2 Biodiesel

*Biodiesel* is defined as the monoalkyl esters of long-chain fatty acids derived from renewable lipid sources. Biodiesel is registered with the EPA as a pure fuel or as a fuel additive and is a legal fuel for commerce. Pure biodiesel has extremely low sulfur content (maximum 50 ppm) and no aromatic content. The cetane number of biodiesel is comparable to that of No. 2 diesel. Since biodiesel is oxygenated (esters contain oxygen), its combustion in diesel engines is more complete than that of petroleum fuels.

The use of biodiesel instead of regular diesel fuel in a conventional diesel engine may result in a substantial reduction of unburned HC, CO, and PM. A slight increase in the $NO_x$ emissions (caused by a significant increase in $NO_2$) was observed for neat biodiesel or biodiesel blends compared with regular diesel fuel [Sharp 1998]. Durbin et al. [2000] observed that 100% biodiesel and biodiesel blends produce slightly higher PM emissions from 1995 Ford 350 than the California 330-ppm sulfur fuel. Durbin et al. also observed significant difference in the fuel effects on emissions for different vehicles. Absence of sulfur also enables the use of catalytic oxidation technologies without the concern over catalyst poisoning or PM emission penalties from the catalytic creation of sulfates. A U.S. Bureau of Mines study reported by Howell and Weber [1997] showed PM reductions of 50% when neat biodiesel was used instead of regular diesel fuel. The test, performed with biodiesel and biodiesel blends in underground mines, also resulted in noticeably less offensive exhaust odor. A study conducted for DEEP by the combined staff of NIOSH, University of Minnesota, Michigan Technological University, and ORTEC reported by Watts et al. [1998] and Bagley and Gratz [1998] in an isolated zone of an underground metal mine compared standard low-sulfur No. 2 diesel fuel with a blend of 55.6 vol % soy methyl ester and D2 fuel. The test vehicle was equipped with a DOC for both fuels. The observed reduction in DPM was approximately 20% when measured by RCD or NIOSH 5040 methods [Watts et al. 1998]. Bagley and Gratz [1998] reported a reduction in solid particle fraction (SOL) of 20% and a reduction of 75% in mutagenic activity.

The synthetic diesel and biodiesel fuels can be used in existing engines and fuel injection systems without negatively impacting operating performance. Additionally, results of tests on Jet A-1 fuel conducted at Southwest Research Institute concluded that biodiesel shows significant lubricity improvement compared to diesel fuel [Howell and Weber 1997]. In general, biodiesel fuel produces lower CO, HC, and carbon particles, but increased soluble OC, so that TPM mass may either decrease slightly or increase.

McCormick et al. [2001] studied the impact of biodiesel chemical structure, specifically fatty acid chain length and a number of double bonds, on emissions of $NO_x$ and PM. Seven biodiesel fuels produced from real-world feedstocks and 14 produced from pure fatty acids were tested in a heavy-duty truck engine using the U.S. heavy-duty Federal test procedure. They found that $NO_x$ emissions increased with increasing fuel density or decreasing fuel cetane number of biodiesel fuel. For tested biodiesel fuels with density <0.89, PM emissions were found to be constant. McCormick et al. also concluded that PM emissions were impacted only when cetane number values were less than those of conventional diesel fuels today. PM reductions were found to be proportional to the fuel oxygen content for biodiesel fuels with cetane number greater than about 45 or density less than 0.89. An increase in $NO_x$ emissions over petroleum diesel was evident, but they could not explain a mechanism that would cause it.

Biodiesel over time will soften and degrade certain types of elastomers and natural rubber compounds. Therefore, precautions are needed when using high-percentage blends to ensure that the existing fueling system, primarily its fuel hoses and fuel pump seals, does not contain elastomer compounds incompatible with biodiesel. If a vehicle's fuel system contains these materials, their replacement with biodiesel-compatible elastomers such as Viton B is recommended. The recent switch to low-sulfur diesel fuel has caused most original equipment manufacturers (OEM) to switch to components suitable for use with biodiesel, but users should contact their OEM for specific information. Fuel-injection equipment manufacturers have agreed that fuels containing up to 5% of biodiesel are compatible with existing equipment.

Petkewich [2001] reported that the first two public filling stations offering biodiesel fuel in the United States were opened in San Francisco, CA, and Sparks, NV, in May 2001. As of November 19, 2001, the price of neat biodiesel is about $2.20. The "Biodiesel Fuel" section of the Technology Guide [DieselNet 2001] provides a comprehensive discussion of this subject. An extensive list of the published literature on biodiesel fuels is available on the EPA Web site [EPA 2001] and on the Web site of the National Biodiesel Board [2001].

### 2.3.2.3 Fuel-Water Emulsions

The potential for reducing diesel emissions by adding water to diesel fuel (fuel-water emulsions) has been extensively explored recently. Introducing water to the combustion chamber of diesel engines has the effect of reducing combustion temperature and thus reducing the production of $NO_x$. Other effects are the reduction of PM and an increase in CO and HC emissions. Fuel-water emulsions have been reported to reduce *both* $NO_x$ and PM by 40% to 50% [DieselNet 1998b; Langer and Daly 1999]. Langer and Daly [1999] also indicated that there was no fuel cost penalty. The additional CO and HC produced can be handled by a DOCC. Special blending technologies, usually involving additives, are required to keep the water and petroleum-based fuel oil together in a stable emulsion. On the negative side, fuel oil-water emulsions suffer from the potential for corrosion of engine parts, freezing, emulsion instability in storage, and reduced lubricity. Ongoing research is addressing these problems and is driven simply by the cost-effectiveness of this technology.

Several low-emission, diesel fuel-water blends and blending technologies are available on the market. For example, A-55, Inc., has developed a clean fuel composed of 20% to 30% water,

70% to 80% petroleum, and 0.5% of the A-55 additive. The A-55 fuel is injected as small droplets into the engine's combustion chamber in the same way as traditional fuels. However, once inside the combustion chamber, the water in the A-55 fuel vaporizes, shattering the fuel droplets into much smaller droplets. This secondary atomization of the fuel results in more complete combustion, reducing particulate emissions that are the product of incomplete combustion.

Lubrizol Corp., in conjunction with Caterpillar, Inc., developed the water emulsion fuel technology named "PuriNOx." The fuel emulsion is applicable to direct-injection heavy-duty diesel engines. According to Lubrizol, the technology is compatible with existing engines and aftertreatment devices. The system requires a relatively elaborate fuel mixing plant and thus is well suited for larger mines with centrally fueled fleets. The manufacturers claim reductions in $NO_x$ emissions from diesel engines of up to 30% and reductions of PM emissions of up to 50%. Using emulsions results in a 10% to 15% loss in engine rated power observed with regular diesel fuel [Lubrizol 1999].

According to the manufacturers, the fuel-water emulsions should be similar to or lower in price than diesel fuels and can be used as ordinary fuel without any modifications whatsoever.

### 2.3.3 Fuel Additives

Metals (such as platinum (Pt), strontium (Sr), copper (Cu), and iron (Fe)) and the rare earths (such as cerium (Ce)), when added to diesel fuel in small concentrations, were found to be efficient at oxidizing the soot, thereby reducing visible smoke [Howard and Kausch 1980]. Test results also showed that fuel additives may decrease the solid PM in raw exhaust by 15% to 25% [Lepperhoff et al. 1995; Mayer et al. 1999]. More significantly, FBCs lower the regeneration temperatures of DPFs. Spontaneous regeneration of uncatalyzed filters without FBC occurs at 550 °C to 650 °C. Use of a catalytic coating on the filter element and an FBC significantly lowers regeneration temperatures. Also, the type and dosage level of the fuel-borne additive make a significant difference in regenerating temperatures. For instance, when a Ce-based FBC is used, continuous regeneration of the catalyzed filters occurs at exhaust gas temperatures >400 °C, while stochastic (balanced or equilibrium) regeneration occurs between 200 °C and 400 °C [Lepperhoff et al.1995; Bach et al. 1998]. Regeneration is unlikely to occur at temperatures <200 °C. Jelles et al. [1999] found that a Pt-Ce fuel additive supports continuous regeneration of a CDPF at temperatures >327 °C.

Today, it is recognized that the primary reason for using an FBC in underground operations is to attain spontaneous regeneration of DPFs at the lowest possible exhaust temperature. Use of an FBC for this purpose is pervasive in the tunneling equipment equipped with DPFs in Europe [Schnakenberg 1999a]. When the FBC is used in combination with a DPF, there is no concern with the toxicity of the metallic ash (emitted as nanoparticles). Furthermore, as noted above, some FBCs, notably the Pt-Ce FBC, are effective at extremely low dosing levels. In fact, Duffy and Samarchi [1997] observed no significant increase in toxicity of the raw exhaust with the optimum dose level of 7 ppm Pt-Ce. Thus, with a DPF in place, the toxic potential of this FBC is insignificant.

Dosage of the FBC is an important subject of optimization. It was found that regeneration quality is not further improved if a threshold dosage is exceeded [Burtsher et al. 1999]. Minimizing the FBC dosage is important in order to minimize costs, to lower the potential increase in exhaust toxicity owing to the emission of the FBC metallic ash in the form of nanoparticles, and to reduce ash buildup on the DPF. A typical dosing level for the Fe-based additive Satacen® is about

36 ppm Fe. The typical dosing of Octel Octimax 4800 is 20/5 ppm of Fe/Sr. Considerably less ash is produced by Pt-Ce because the dosing level is much lower (2-8 ppm, with the Pt usually <0.5 ppm). The accumulated ash will eventually (at >2,000 hr of operation with Satacen and proportionally longer for Pt-Ce) require cleaning the DPF.

In addition to the minor problem of ash accumulation, a more serious concern related to the use of the fuel additives is the emission of the FBC metal oxide upon DPF malfunction and the unintentional use of fuel containing an FBC in a vehicle not equipped with a DPF. Periodic monitoring of DPF performance could prevent potential for extended exposures to FBC ash. It is important to ensure that fuel containing the FBC is used only by engines with DPFs. Furthermore, adding the FBC to the equipment fuel tank upon fill-up does not ensure proper dosage. On-board fuel-dosing systems are available on the market, but they are not yet perfected [Schnakenberg 1999a]. For these reasons, it is necessary to use separate fuel storage and handling systems at this time. One system stores and dispenses fuel for vehicles without DPFs, the other stores and dispenses fuel to which an FBC has been added and is intended for use by vehicles with DPFs. This segregation of fuel is the only burden to the mine imposed by FBC use. When perfected, the on-board dosing systems would eliminate the need to segregate fuel supplies for DPF-equipped and non-DPF-equipped vehicles.

The selection and formulation of the FBC is probably the domain of the fuel additive manufacturer and the DPF system supplier. Temperature-time profiles of the engine exhaust temperature, taken during full-shift operation of the vehicle intended to receive a DPF, are essential in selecting an FBC. Some of the FBC products on the market are:

(1) EOLYS-DPX9® (Ce) by Rhodia (formerly Rhone-Poulenc, Inc.);
(2) Platinum Plus® (Pt-Ce) by Clean Diesel Technology (a Ce-based additive supplied by Rhodia);
(3) Ferrocene (Fe) by Aldrich; Satacene®, Sat Chemie Gmbh, now Octel, U.K.;
(4) OS-96401(Cu) by Lubrizol; and
(5) Octimax® OCI-4800, Fe-Sr; Octel America, Inc.

The Health Effects Institute (HEI) conducted a study [HEI 2001] on health risks associated with Ce-based additives and concluded that using Ce as an FBC with a particulate filter would result in a measurable increase in the ambient levels of cerium oxide in particles <0.5 m (perhaps up to several orders of magnitude greater than current levels) depending on the level of Ce actually used, the filter efficiency in trapping the particles, and the degree of penetration in the vehicle fleet. HEI found as result of short-term diesel engine tests that despite the high efficiency of filters in trapping PM (>90%), a small amount of Ce was emitted in the particulate phase of the exhaust. Ce mass relative to the total particle mass was found to be between 3% and 18% based on two tests using two different types of filters. Based on the limited data available, HEI found that toxicity of cerium oxide seems to be small and that cerium oxide might not be of concern when inhaled at the low levels.

### 2.3.3.1 Review of the Published Research

Lepperhoff et al. [1995] conducted a study on the performance of a Ce-based fuel additive (DPX6, Rhone-Poulenc) in relation to particulate trap (Corning EX-47) regeneration quality, trap filtration efficiency, particle size distribution, and fate of the additive under steady-state engine operating conditions. Lepperhoff et al. found that a Ce FBC, when added at a dosage of 50 ppm of Ce by weight, reduced particulate emissions by roughly 20%. The DPM reduction mainly resulted from the reduction of the EC content, and the quantity of the volatile HC was unaffected. In addition, the tested fuel additive lowered the collected PM ignition temperature to 200 °C. When using the Ce fuel additive at the Ce dosage of 50 ppm by weight and the Corning EX-47 trap, DPM emissions were reduced by >90%. Ninety-seven percent of the Ce compounds were filtered by the trap.

Jelles et al. [1999] examined the performance of different additives. The results of their testing are shown in table 5.

**Table 5.—Minimum temperature for continuous regeneration** [Jelles et al. 1999]

| Additive | Concentration, ppm wt. | Filter | Minimum temperature, °C |
|---|---|---|---|
| None | — | EX-80 | 540 - 560 |
| None | — | Pt EX-80 | 420 - 430 |
| Ce | 100 | EX-80 | 430 |
| Pt-Ce | 0.5 - 5 | Pt EX-80 | 330 |
| Pt-Cu | 0.5 - 5 | Pt EX-80 | 350 |
| Pt-Fe | 0.5 - 22 | Pt EX-80 | 360 |

NOTE: EX-80 designates the type of Corning monolith filter material; Pt EX-80 indicates that the substrate is catalyzed with Pt.

The highest temperature observed during a regeneration of the filter was >900 °C.

Mayer et al. [1999] found that the use of Cu-based fuel additives resulted in elevated dioxin and furane emissions. Therefore, Mayer et al. suggested that Cu additives should not be used in the fuel for underground machinery. The tests also showed that the use of Fe and Ce additives did not result in elevated dioxin and furane emissions.

Burtsher et al. [1999] examined the effects of the use of Ce-based fuel additives on the emission of nanoparticles and the concentration of the additive in the exhaust. The measurements were made on the exhaust generated by a heavy-duty diesel engine (Liebherr 914T) and a small naturally aspirated Yamaha diesel engine. The Ce additive at concentrations of 20 and 100 ppm was found in the particulate phase of the exhaust. The measurements taken by the scanning mobility particle sizer (SMPS) showed that the concentrations of nanoparticles significantly increased with increases in the concentrations of the additive in the fuel. Chemical analysis showed that the small particles consisted only of the additive material. Since there is concern with human exposure to nanoparticles, Burtsher et al. concluded that diesel exhaust filters should be used to eliminate emission of additive-based nanoparticles.

In sum, FBCs should be used mainly to enhance the regeneration performance of DPFs. Care should be taken to ensure that FBCs are added only to the fuel that will be used exclusively by engines equipped with the DPFs. Filters successfully curtail emissions of the metallic ash resulting from the use of additives. It is equally important to avoid use of the regular fuel with filter-equipped engines. With regular fuel, the filter may fail to regenerate, consequently overloading the filter with soot and eventually causing the filter and engine to malfunction. The additional cost associated with the use of an additive is usually less than 10 cents/gal (4 to 7 cents/gal for Platinum Plus®, for example). MSHA requires [30 CFR 75.1901(c)(1996)] that any fuel additive used in underground coal mines be registered with the EPA in accordance with 40 CFR 79. According to the literature, most of the commercially available additives are EPA-approved; all except the Cu-based additives are acceptable for underground use.

## 2.4 Aftertreatment Technologies

### 2.4.1 Diesel Oxidation Catalytic Converters (DOCCs)

The primary function of a DOC is to oxidize CO and HC to $CO_2$ and water in an exhaust gas stream. The role of the catalyst is to increase the oxidation rate without itself being consumed in the process. The catalyst also substantially reduces the temperature needed for oxidation of the CO and HCs. Catalysts are characterized by their activity and selectivity. Both characteristics are influenced by temperature. DOCs reduce DPM emissions by oxidizing some of the less volatile HCs that contribute to the SOF of the PM mass or become bound to the soot particles by adsorption. On the other hand, DOCs have no effect on the solid core carbon particles (soot) that also make up DPM. The efficiency of a DOC can be reduced by catalyst poisoning or excessive accumulation of the DPM on the catalyst's surfaces. A catalyst can be poisoned by fuel sulfur and compounds from lubrication oils.

Some other gaseous components of diesel exhaust are not fully oxidized and thus are also candidates for oxidation by the DOC. NO and $SO_2$ are of particular concern. Nitrogen and oxygen (air) spontaneously combine at high combustion temperatures to form mostly NO and a little $NO_2$. $SO_2$ is produced by oxidation of the sulfur in the fuel during combustion. Further oxidation of NO and $SO_2$ takes place in the DOC at high exhaust temperatures to produce sulfate and $NO_2$. The unfavorable gaseous phase reactions taking place in the DOC are $NO \rightarrow NO_2$ and $SO_2 \rightarrow SO_3$. These reactions increase the toxicity of the emitted exhaust. NO, with a TLV of 25 ppm and a toxicological behavior similar to that of CO, is converted to the acid gas $NO_2$, which has a ceiling of 5 ppm and attacks the mucous membranes, increasing the likelihood of infection and causing long-term effects from constant irritation as well. The impact of an increase in $NO_2$ emissions on local workplace concentrations depends on the distance from the tailpipe and whether there is a mechanism for adsorption of $NO_2$ (such as the rock dust and wet walls in coal mines). $SO_2$ is also converted to $SO_3$, which further combines with water to form sulfuric acid aerosols. The TLV for sulfuric acid is 2.5 mg/m$^3$. It is important in the purchase of DOC to specify that it is for diesel use, the fuel sulfur level, and that NO conversion should be specified if not controlled.

Performance of a DOC is a function of the fuel composition and exhaust temperature. When standard, low-sulfur, Federal diesel No. 2 fuel (D2) containing an average of 340 ppm by weight of sulfur is used, the exhaust exiting the DOC may contain elevated concentrations of sulfuric acid and sulfate aerosol. Researchers [Mayer 1997; Klein et al. 1998; Carder 1999] observed substantial increases in particulate mass and particle number in exhausts treated with DOCs. They noticed that, depending on the concentration of sulfur in the fuel and the exhaust temperature, DOCs may enhance the emission of sulfate aerosols. The increase in PM mass and number was, however, more pronounced for those engine operating conditions that generate high temperature exhaust and thus provide favorable conditions for sulfate formation. Therefore, the sulfur content of diesel fuel is critical in the design and application of catalyst technology. The use of ULSFs (<50 ppm by weight) results in substantial reduction in sulfate formation and TPM emissions. The use of low-sulfur fuel also reduces the risk of poisoning a catalyst. Since a catalyst is more efficient at high temperatures, a DOCC needs to be positioned close to the engine. A heat retention blanket should be used in the case of longer exhaust runs to maintain exhaust temperature.

The science of formulating exhaust catalysts involves numerous factors, including selection of metal or metal combinations, supporting structure, interactions with stabilizers and promoters, and heat treatments. Catalyst formulations are usually proprietary. DOCCs presently in use mostly appear as two forms of cellular monoliths [DieselNet 1998a] that have replaced the pelletized forms. Figures 1 and 2 illustrate the differences between ceramic monolith [DieselNet 1997a] and the metallic monolith [DieselNet 1997b] design. Both are characterized by a high substrate area-to-volume ratio. They are usually enclosed in a stainless steel container adapted to fit the exhaust system. The size of the DOCC needs to be optimized with respect to engine size. The volume of the DOCC should be approximately equal to engine displacement [MECA 1999a].

DOCCs are extensively used for on- and off-road applications. They are also used by underground mine operators as an emission control device to reduce odor, HC, and CO emissions from diesel equipment [McClure et al. 1988]. Mayer [1997] has suggested that the positive effects of using DOCCs are irrelevant to construction site diesel engines used in tunnels. Therefore, a DOCC should not be deployed for utility vehicle diesel engines in an underground environment because the negative effects far outweigh the benefits. However, if and when ULSF (<50 ppm by weight) or sulfur-free fuel is available and used, DOCs with formulations that minimize the formation of $NO_2$ can be used to achieve significant reductions in CO and HCs (and thus DPM and odors).

Catalyst substrates are designed to last the entire lifespan of the engine. The substrates can stand harsh operating conditions and offer good thermal durability. The major reason for DOCC failure is deactivation of the catalyst. Catalysts can be deactivated by high temperatures (>650 °C) and poisoned by lubricating oil additives (phosphorus, zinc, heavy metals) and fuel sulfur [DieselNet 1998c]. Leaks of lubricating oil into the exhaust system are very detrimental to catalyst life. Heywood [1988] and the more current "Technology Guide" at www.DieselNet.com/tg.html provide good reviews of DOC and DOCC technology.

#### 2.4.1.1 Review of Published Results

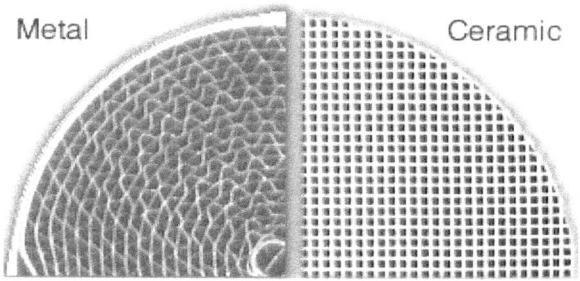

Figure 1.—Monolithic catalyst substrates.

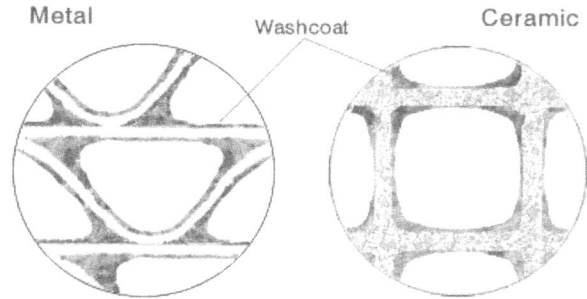

Figure 2.—Catalyst washcoat.

Because catalyst formulations can be varied to accommodate various objectives and fuel sulfur levels, the performances reported must be carefully scrutinized. Early studies such as McClure et al. [1988], which investigated the effectiveness of DOCCs used in an underground mine, found that although DOCCs are effective at reducing CO, HC, and odors when the exhaust temperature remains >250 °C, they increased $NO_x$ emissions slightly. Sulfate emissions were found to increase, but the fuel sulfur level was >1,000 ppm, whereas fuel sulfur levels for fuel currently available in the United States averages 350 ppm.

Pataky et al. [1994] investigated the effects of a DOCC on regulated and unregulated emissions from a 1991 prototype Cummins L10-310 diesel engine fueled with 100 ppm by weight sulfur fuel. The DOCC with metallic substrate and Pt coating was supplied by the Degussa Corp. Pataky et al. reported that the DOCC had no significant effect on $NO_x$ and NO at any test mode. The DOCC reduced HC emissions by 60% to 70% and TPM by 27% to 54%, primarily as a result of a 53% to 71% reduction of the SOF. The DOCC increased $SO_4^=$ at the higher temperature modes, but had no effects at the lowest temperature mode.

DOCCs were used in the DEEP biodiesel study [Watts et al. 1995; Bagley and Gratz 1998]. Tailpipe emissions under torque converter loading showed 98% to 99% reduction in CO, but an increase of $NO_2$ by 185% with the standard low-sulfur D2 fuel and 233% with the 55.6% blend of biodiesel and the standard fuel. This study also included 1 day of running without a DOCC (compared to 4 days of testing each on the blend and straight D2). Without the DOCC, CO was significantly greater than CO with the DOCC; the $NO_2$ was about one-third of the value with the DOCC.

Carder [1999] reported that the tested DOCC reduced HC and CO by an average of 72% and 93%, respectively, while $NO_x$ emissions were not significantly affected. Interestingly, the exhaust treated in the DOCC contained on average 66% higher DPM. The substantial increase in DPM emissions was attributed to sulfate formation.

Six different DOC formulations were tested on a Detroit Diesel Series 60 engine by MECA [1999a]: two of low-activity, two of medium-activity, and two of high-activity. The fuel sulfur level was 368 ppm for the baseline. MECA researchers found that an optimized catalyst system can achieve emission reductions of >35% for PM and 70% for HC and CO. Gaseous emission reductions of HC and CO were found to be directly related to catalyst activity. The researchers found that DOCs are extremely effective at reducing PAHs and other HC emissions. They also found that a DOC can be used in conjunction with an FBC to offset increased PM emissions resulting from the use of EGR.

In conclusion, DOCCs are formulated specifically for diesel engine use and to balance the beneficial reductions of CO and HC with the increase in $NO_2$ and sulfates. Use of ULSFs allows catalyst formulations that are more effective in reducing CO and HC without the penalty of sulfate (counted as DPM mass) formation. Conversion of NO to $NO_2$ may be an issue.

### 2.4.2 Diesel Particulate Filters (DPFs)

DPFs were found to be technically feasible, controllable in the field, and a cost-effective technology for controlling DPM in European tunneling work [Mayer 1997] and Canadian underground metal mines [McGinn 2001a]. Two performance aspects of DPFs are crucial: the filtration efficiency of the system and the ability of the system to regenerate and provide long-term operation without diminishing the filtration efficiency of the filter and performance of the engine. The design and performance of DPF systems strongly depend on the vehicle/engine duty cycle, and the systems require optimization for specific applications. Vehicle/engine type and operating conditions must be recognized before the design and optimization process.

#### 2.4.2.1 Particulate Filter Design

*Filter Media*

Filters control DPM emissions by physically trapping soot particles in their structure. Major designs of particulate filters on the market are based on media such as porous cordierite and silicon carbide (SiC) wall-flow monoliths and on deep-bed fiber filters constructed from matted, woven, or knitted glass or ceramic fibers. Wire mesh, nonwoven SiC ceramic fiber, and sintered metal substrates are some of the alternative substrates still under scrutiny by researchers. Filter media can be catalyzed to enhance filtration efficiency and removal of some gaseous compounds and to lower the regeneration temperature.

Wall-flow monoliths, namely the cordierite filters from Corning, Inc., and NGK Insulators Ltd., are the best known and have the longest history of use (since the early 1980s) among all available materials (figure 3). Exhaust flows through the porous ceramic walls (wall flow), and DPM is collected on the upstream side of the walls, as shown in figure 4. The surface of the upstream wall may also contain a catalytic washcoat to help lower the autoregeneration temperature. Corning's EX-80 is currently the most popular wall-flow monolith material. Recently, Corning developed a new material (DuraTrap RC) with larger filtration areas and better thermal and mechanical properties [Corning, Inc. 2000].

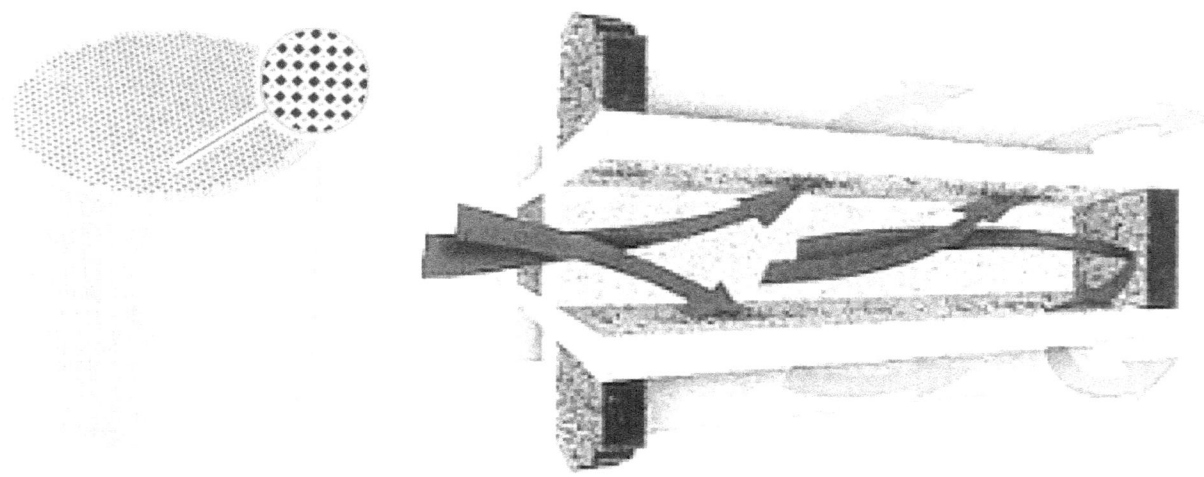

Figure 3.—Example of a ceramic monolith filter element.

Figure 4.—Gas flow in a monolith, wall-flow filter. (Photograph courtesy of Corning Incorporated.)

In recent years, SiC materials (Ibiden Co. Ltd., NGK, NoTox A/S) has been successfully established as a viable alternative to cordierite. SiC has lower thermal shock resistance, but its higher melting temperature makes it more durable over cordierite substrates during uncontrolled regeneration. SiC filters also have higher porosity than cordierite, which results in less back pressure. SiC filters are generally more expensive than equivalent filters made of cordierite.

Fiber filters are classified as deep-bed or depth filters because the particles are trapped deep within and onto the filter fibers directly, as shown in figure 5. The filter material is supported between two long, narrow concentric cylindrical grids or mesh to form a thick-walled tube. The exhaust flows through the walls of the tube, and several of these tubes are assembled to form a filter (figure 6). The fiber surfaces of these filters may also be catalyzed.

Mayer et al. [1995] found that the ceramic monolith surface filter and the deep-bed filter of knitted fibers have distinctively different properties. When evaluated on a gravimetric basis, both systems showed comparable efficiencies of around 90%. When evaluated on a particle-count basis, the efficiency of the surface filter was <70%, while that of the deep-bed filter was >90%. The efficiency of the surface filters was found to deteriorate for particles <100 nm, falling practically to zero for the 30-nm-diam particles. The efficiency of the surface filter increased with loading (formation of the filter cake), and there was a simultaneous progressive increase in back pressure. The filtration efficiency of knitted fiber filters was found to be highest in the new state, but deteriorated slightly with increased loading.

The efficiency of a DPF in the removal of CO, HCs, and the OC fraction of DPM from the exhaust stream can be enhanced by adding an oxidation catalyst to the filter material. In addition, a catalyst lowers the ignition temperature for initiating the autoregeneration process. DPFs can be catalyzed by a washcoat or by deposition of catalytic material from an FBC. The formulation and quantity of a catalyst need to be designed and optimized for a particular application, specific to the exhaust temperature of the engine considered, the engine duty cycle, and the formulation of the fuel. Armed with these parameters, the DPF manufacturer can specify the appropriate catalyst technology or recommend other means to accomplish DPF regeneration. An excess of catalyst in a system may result in increased emissions of sulfates or $NO_2$ and increased toxicity of the exhaust. An insufficient amount of catalyst in a system may result in reduced efficiency of the system and/or the inability of the system to regenerate under the duty cycle. Properly optimized amounts of a catalytic coating and FBC allow for the use of Federal diesel No. 2 fuel with its specified maximum of 500-ppm sulfur.

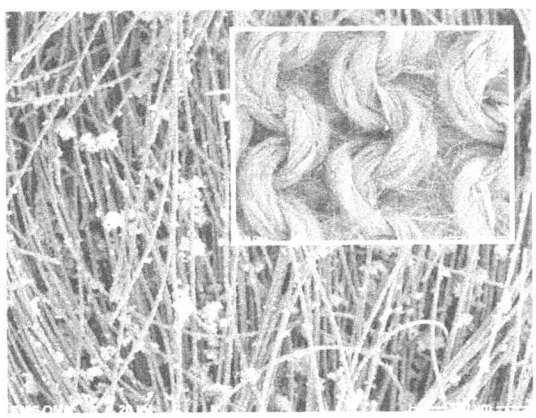

Figure 5.—Knitted microfiber filter (deep-bed filter).

Figure 6.—Example of a particulate filter system using fiber media.

DPF size depends on the size of the engine, the engine-out DPM emissions level, and the regeneration technique used. When systems are not capable of autoregeneration, filters are significantly larger to accommodate soot collected during relatively long periods of operation without imposing excessive exhaust back pressures.

Installations of a DPF system require instrumentation for monitoring the exhaust gas back pressure. A gauge indicating exhaust back pressure as well as a visual and audible alarm should be installed on the dashboard of a vehicle where the driver can observe it to ensure that the back pressure remains under limits recommended by the engine manufacturer (usually between 100 and 200 mbar). Thus, the driver can take corrective action if the back pressure exceeds recommended limits. Postnikoff [1999] found that the back pressure gauge, more specifically the tube from the exhaust pipe to the gauge, is the weakest component of a DPF system. This tube may become clogged with soot during use and needs to be cleaned periodically using compressed air.

Maintenance of an autoregenerating DPF consists of, at most, removing it from the vehicle and cleaning accumulated ash from it. The need for these actions can be detected by an increase in baseline back pressure. The need for cleaning usually occurs at 2,000 hr or more depending on the FBC dosing and amount of lubrication oil ash. This period can be extended by using lubrication oils with low ash content and, when needed, fuel catalysts that require low dose rates. Additional efforts are needed to support filters that require off-shift regeneration. In addition to routine (monthly) maintenance checks on vehicle emissions is the need to measure the exhaust DPM to verify performance. According to the experiences of manufacturers, researchers, and operators, DPF systems do not seem to cause any additional engine wear or otherwise affect vehicle maintenance.

Approximately 1,600 knitted glass fiber DPFs manufactured by Oberland Mangold GmbH have been deployed on different diesel engines, mostly in tunneling and mining [Kahlert 1999]. The exhaust temperatures are high enough to cause the collected PM to be removed by oxidation sometimes with the assistance of an FBC. Some of the Oberland Mangold systems have achieved over 8,000 hr of operation without failure or loss of efficiency. Postnikoff [1999] reported that a typical service life for a ceramic DPF deployed on vehicles in Agrium, Inc., potash mines is 5 years under severe service. Significantly, the author also reported that a few of the Engelhard Corp. units (ceramic DPF) presently have 10 years of service. Early on, premature failures of the ceramic DPF also occurred and were attributed to extreme vibrations and shock or improper canning.

*Regeneration Methods for Particulate Filters*

The soot collected by DPFs needs to be removed to avoid excessive fuel penalty and damage to the engine and the filter. The removal of the soot, termed "regeneration," is a rather complex process. Many process parameters must coincide to ensure regeneration that does not harm the filter. The governing parameters are exhaust gas temperature, exhaust gas back pressure, the remaining oxygen content in the exhaust gas, volumetric flow rate, etc. When the temperature in a DPF exceeds the required soot ignition temperature, the DPM burns and the back pressure decreases.

Depending on the exhaust temperatures, DPF systems are designed to regenerate on-board the vehicle during on-shift use, or either on-board or off-board while the vehicle is off-shift. Regeneration must occur at intervals that are frequent enough to ensure that the filters do not become overloaded. The concept of on-board, on-shift regeneration was found to be superior to off-shift regeneration due to significantly lower operating costs. Most importantly, on-shift in-line regeneration systems offer unrestricted vehicle operation during the regeneration process and are thus favored by vehicle/engine operators.

The temperatures at which DPM burning occurs are generally higher than the exhaust temperatures commonly achieved by modern engines. Tests by Bach et al. [1998] showed that uncatalyzed filters spontaneously regenerate at 550 °C to 650 °C. However, the exhaust temperatures in most heavy-duty diesel applications do not exceed 450 °C. Modern turbocharged diesel engines at low loads run at even lower exhaust gas temperatures, very often <200 °C [Bach et al. 1998].

A number of techniques have been developed for active and passive on-board regeneration of filters during continuous operation of a vehicle. The passive approach requires that vehicles operate at high part-loads or at full loads for at least 20% to 25% of the time with idling periods minimized. The deployment of particulate filters relying on passive regeneration is not recommended if the engine/vehicle is operating exclusively at low to medium part-load.

The active regeneration approach allows more flexibility regarding engine operating conditions, but requires a source of energy for the heaters. Thus, it is imperative when considering using a DPF to obtain a temperature-time profile of the exhaust temperature for the vehicle to be equipped with the DPF. Armed with the temperature profile, the DPF manufacturer can specify the appropriate catalyst technology or recommend other means to accomplish DPF regeneration.

Field data indicate that mine vehicles, depending on the type of operation, spend a significant percentage of time at engine operating conditions that do not favor passive regeneration. Data obtained from the engine control management system of trucks at Noranda, Inc.'s Brunswick Mine [McGinn 2001b] show that those vehicles average over 30% of the time at low idle.

During continuous regeneration, the average rate of PM mass accumulation and the average rate of removal are in balance. At low exhaust temperatures, mass accumulates until the rate of accumulation equals the rate of removal by burning and an equilibrium loading and back pressure are reached. At these conditions, the engine back pressure should remain within acceptable limits. If regeneration does not occur frequently enough, a DPF may become overloaded with DPM. Excessive accumulation of DPM may result in uncontrolled regeneration. During uncontrolled regeneration DPM burns too quickly; this results in extremely high temperatures and CO emissions. Regeneration can cause high thermal stresses in the filter, which leads to cracks in the material. In addition, at high regeneration temperatures, chemical reactions can occur when fuel additives are present. This might result in changing the crystal structure, strength, and filtration properties of the ceramic filter.

Passive regeneration of a filter is promoted by use of catalytic coatings and/or FBCs. Base metals (Fe, Sr), precious metals (Pt), and rare earths (Ce) have roles in reducing the ignition temperature necessary for oxidation of the PM. To obtain continuous regeneration, a DPF needs to reach the regeneration temperature frequently during a vehicle/engine duty cycle. The dosage of the coating and FBC needs to be optimized with respect to vehicle/engine operating conditions and fuel type used. In experiments, unsuccessful coatings on filters were found to reduce particle collection efficiency and reduce aerodynamic regeneration effectiveness of DPFs [Larsen et al. 1999]. Excess catalyst may also result in increased emissions of sulfuric acid [Carder 1999] and $NO_2$.

Some of the engines deployed in underground mines are not candidates for passive, catalytic, on-board regeneration. Postnikoff [1999] found that most of the medium-duty outby vehicles in the Agrium potash mines do not work hard enough to meet the minimum 350 °C required for deployment of catalyzed ceramic DPFs. Active regeneration techniques are the only option for continuous on-board regeneration when an engine/vehicle operates exclusively or mostly at low to medium part-load and thus exhaust temperatures are too low for obtaining passive regeneration. In active regeneration methods, the exhaust gas, and thus the filter, is heated to the necessary oxidation temperature by an external energy source. Under such a system, the filter collects the particulate contained in the exhaust gas during the so-called loading phase. The particulate load is burned intermittently, during the regeneration phase, when the entire exhaust gas is heated up to a temperature level such that the accumulated PM will begin burning.

Electric heaters (DCL, Engelhard, ECS, etc.) and diesel fuel burners (Deutz) are some of the most common techniques used for active regeneration. Diesel fuel burner technology is well established and safe for DPF regeneration and is primarily designed for city bus application. This technology is well established in Europe, but Deutz, so far, has not introduced this technology to the North American market. The regeneration process for such a system takes approximately 10 min [Houben et al. 1994]. Required control and regulation instrumentation results in substantial initial costs. In addition, a moderate fuel penalty is associated with the fuel burner system.

Necessary regeneration temperatures can also be obtained by heating the exhaust gas or the trap material by means of electrical heaters. Due to high energy requirements, electrical heating has mostly been used for stationary regeneration of DPFs. However, some engine and aftertreatment equipment manufacturers are developing alternative techniques for heating traps, such as microwave technology [Popuri et al. 1999]. These active trap regeneration systems are generally expensive and energy-intensive. Therefore, manufacturers are working on solving these issues that currently are substantially limiting the scope of the application of such systems.

Passive and active technologies can be combined to enable trap regeneration to take place at low exhaust gas temperatures, which reduces demand for external energy and lowers the cost for the active components of the system. Bach et al. [1998] tested a particulate trap regeneration system that combined the advantages of an FBC (in this case Ce) and additional electric heating. Bach et al. found that trap regeneration can take place at low exhaust gas temperatures of about 270 °C. The consumption of electrical energy was significantly reduced due to the catalytic action of the additive.

The alternative to on-board regeneration during operation is to perform either an on-board or off-board regeneration while the vehicle is off-shift. On-board regeneration is usually performed by means of electrical heaters integral to the DPF and an off-board control unit. Off-board regeneration requires removal of the DPF and replacing it with a regenerated unit. The soot-laden filter is placed in a kiln where it is heated under controlled conditions. Either of these two regeneration procedures must usually be performed after every shift; thus, a vehicle is immobilized during regeneration. Manufacturers are working on shortening the time necessary for the regeneration process. Unikat AB has developed a system that can be regenerated in as little as 30 min [Unikat AB 1999]. According to the manufacturer, the off-board regeneration of the Engelhard SPX soot filter requires only 14 min [Engelhard Corp. 1999]. A DCL filtration system retrofitted to a load-haul-dump vehicle at Noranda's Brunswick Mine uses an onboard electrical heating system as a backup system in cases when the filter does not passively regenerate. The filter system must be connected to the regeneration station for approximately 2 hr to achieve complete regeneration [McGinn 2001a]. An ECS Omega diesel particulate system installed on a light-duty tractor at INCO's Stobie Mine required <10 min at the regeneration station after each shift [Nault 2001].

All filters, whether passively or actively regenerated, must, after a certain number of hours in operation, be cleaned from the accumulated ash. The accumulated ash results in a gradual increase in exhaust back pressure of the regenerated filter. The exhaust back pressure should be regularly monitored, and the filter should be cleaned before back pressure jeopardizes engine or filter performance. The rate of ash accumulation is a function of several factors, including engine condition, primary oil consumption, formulation of oil and fuel, and use of fuel additives containing metals. Due to complexity and requirement for specialized equipment, this cleaning procedure is usually performed by the filter manufacturer.

### 2.4.2.2 Review of Published Results

Baz-Dresch et al. [1993] tested an uncatalyzed Corning EX-66 cordierite filter (the advertized collection efficiency of the uncatalyzed trap was 65% to 70%). Testing was conducted on a Caterpillar 3304 engine, designed for mining applications. The engine was tested under six of the ISO 8178 C1 eight-mode, steady-state operating conditions (I50, I75, I100, R50, R75, R100). The engine was fueled with low-sulfur fuel. Baz-Dresch et al. found that the regeneration temperature increased from 405 °C measured after 839 hr of operation to 450 °C measured after 2,881 hr of operation. Gaseous and DPM emission measurements indicated deterioration of the filter medium over time. Below are the reduction efficiencies obtained after 839, 1,584, and 2,881 operating hr; the ranges result from different behaviors at different engine modes (and exhaust temperatures):

**Reduction efficiencies**

|  | Operating hours | | |
| --- | --- | --- | --- |
|  | 839 hr | 1,584 hr | 2,881 hr |
| CO, ppm | 21.3% to 64.8% | 14.3% to 57.5% | 2.9% to 42.1% |
| HC, ppm | 5.4% to 89.7% | 23.1% to 76.5% | 39.4% to 83.2% |
| DPM (mg/m$^3$) | 48.1% to 94.5% | 41.2% to 81.9% | 28.5% to 82.2% |

Baz-Dresch et al. [1993] speculated that damage to the DPF occurred because of mechanical shocks and vibration, cracking or melting of the ceramic due to uncontrolled regeneration, or a cracked substrate that resulted from high thermal gradients during repeated regeneration.

The U.S. Bureau of Mines conducted a study to evaluate the performance of a CDPF alone and combined with a DOC at two metal mines (Q and T) and under laboratory conditions [Watts et al. 1995]. The tested DPF was a ceramic monolith. Three methods were used to measure DPM and respirable dust concentrations in the two mines: personal diesel exhaust aerosol sampler (PDEAS), micro-orifice uniform-deposit impactor (MOUDI), and RCD sampler. The mean DPM reduction efficiency per unit production for the DOC + CDPF system, installed on a Caterpillar 3306 engine, was $71 \pm 29\%$ for mine Q. The efficiency of the CDPF was estimated to be $72 \pm 21\%$ (PDEAS) or $62 \pm 25\%$ (RCD) for mine T.

Mayer [1997] reported 80% to 85% efficiency for a DPF tested in a study by VERT (Verminderung der Emissionen von Realmaschinen im Tunnelbau). This efficiency was assessed on the basis of "pure" gravimetric PM evaluation. The particulate count reductions were reported to be >90%. The filtration efficiencies for soot were >90%, and the filtration rate for nanoparticles >99%. From extensive testing, Mayer concluded that the DPF using a monolithic element was very efficient at filtering large diesel exhaust particles. However, for this and other DPFs using different filtration media, massive breakthrough of the smallest sized diesel particles—the nanoparticles—was found. This was also observed in DPFs using sintered metal filters. However, for the DPF using the fiber element, the deep-bed filters, the efficiency increased with a decrease in particle size. Mayer concluded that catalyzed diesel particulate traps are extremely efficient and state-of-the-art.

MECA [1999a] found that a DPF, when used with 54-ppm (instead of 368-ppm) sulfur fuel, reduced PM, HC, and CO emissions from a Detroit Diesel Corp. (DDC) Series 60 engine exercised over the U.S. heavy-duty transient FTP cycle by 70%, 94%, and 63%, respectively. When the DPF was used with FBCs, the PM emissions were reduced by 78%. MECA concluded that the PM emission levels of 0.005 g/bhp-hr were achievable by using a DPF with a zero-sulfur fuel.

Larsen et al. [1999] tested three different types of particulate traps: a Corning EX-66 cordierite substrate coated with a micromembrane by CeraMem Corp., an Ibiden SiC material, and an Ibiden SiC substrate coated with a ceramic micromembrane. All of the filters achieved >93% total filtration efficiencies (by mass); the regular SiC filter reached levels of 97%. For the small particles, the uncoated SiC trap performed the best; it reduced the DPM by a factor of 99%.

Carder [1999] reported results of testing a Caterpillar 3306 engine retrofitted with a CDPF designed for that particular engine by Clean Air System. The overall weighted ISO 8178 eight-mode average DPM reduction of the system was 72%. The average reductions in HC, CO, and $CO_2$ were 88%, 83%, and 21%, respectively. Oxides of nitrogen were not substantially reduced.

The DETR/SMMT/CONCAWE Particulate Research Programme [Andersson and Wedekind 2001] has investigated the effects of engine technologies and fuel specifications on regulated PM, particle number, mass, and size. The largest effects of a single technology on particles was observed with DPFs (oxidation catalyst followed by a DPF), where particle mass and number were reduced by several orders of magnitude. The exception was at high exhaust temperature conditions, where significant numbers of nucleation mode particles were emitted after the DPF. The filter reduced integrated particle mass emissions by about 90%. Filter effects dominated within the accumulation mode (particles sizes between 0.08 and 1 $\mu$m).

The harsh environment makes deployment of filters on diesel-powered underground mining equipment a challenging task. So far, limited data are available in the literature on the performance of DPFs deployed on underground mining production equipment. One of the valuable sources of such data is the trap evaluation studies at Noranda's Brunswick Mine and INCO's Stobie Mine. These studies, sponsored by DEEP, have the evaluation of the different DPFs deployed on heavy- and light-duty production vehicles as the long-term goal. The preliminary results of those studies were presented at the Mining Diesel Emissions Conference (MDEC 2001).

## 2.4.3 DOCC and DPF Combinations

Systems incorporating a DOCC and DPF are commercially available; such systems have been tested by several researchers. The combined systems incorporating both a DOCC and DPF were designed to provide good reductions of gaseous and PM emissions.

The CRT developed by HJS and Johnson Matthey is an example of such a system. The CRT system consists of an extremely active DOCC followed by a DPF. CO and HC are almost completely oxidized to $CO_2$ and water. The catalyst was formulated so that it also oxidizes a substantial portion of the NO to $NO_2$. The $NO_2$ generated oxidizes the carbon in the particulate trap, thus performing regeneration at low exhaust temperatures. Trap regeneration occurs at temperatures between 200 °C and 450 °C, low enough to result in almost continuous regeneration. An advantage of continuous regeneration is that the local peak thermal stresses can be avoided. Unfortunately, substantial amounts of $NO_2$ are not utilized and pass through the DPF ($NO_2$ slip). Three key requirements for the system to perform properly are reasonable balance between $NO_x$ and particulate emissions, a duty cycle that regularly gives rise to exhaust gas temperatures >260 °C, and the use of ULSF.

CRT systems have been successfully applied for curtailing DPM emissions from city buses. Czerwinski et al. [1998] tested an HJS-CRT system on a Liebherr D914T construction engine as part of the VERT suitability project. They reported very good filtration efficiency of the system at all test conditions and recommended the system to the users. The average reductions over the ISO 8178 eight-mode test were about 85% for PM, 94% for CO, 39% for HC, and 21% for $NO_x$. Despite decreases in $NO_x$ concentrations, the concentration of $NO_2$ increased when the exhaust was treated in the CRT system. This increase in $NO_2$ has caused a reluctance on the part of the manufacturer to offer systems for use in underground tunneling and mining operations. (This was, in fact, the reason given for their declining to bid on a DEEP DPF evaluation project [Schnakenberg 1999b]).

### 2.4.3.1 Review of Published Results

Czerwinski et al. [1998] tested a HJS-CRT system on a Liebherr D914T construction engine. The HJS-CRT system consists of a ceramic monolith oxidation catalyst in line with a ceramic monolith particulate trap. The engine was fueled by Swiss standard diesel fuel (sulfur 500 ppm, cetane number 48) and Greenenergy ULSF (sulfur 25 ppm, cetane number 50). The engine operating conditions were: rated speed, 100% load (R100); intermediate speed, 100% load (I100); R50, I50, I25, I10, as defined in ISO 8178 [30 CFR 7 (1996)]. The evaluation was performed on the basis of measurements taken by an SMPS, PAS, and ELPI and by gravimetric and coulometric (reports EC) analysis. The observed average filtration efficiency was 85%. The count of nanoparticulates was also reported to be reduced efficiently. The HJS-CRT system also reduced concentrations of CO, HC, and $NO_x$ by 93%, 35%, and 21%, respectively. An increase in $NO_2$ concentration of in the exhaust treated in the system was also reported.

Kauffeldt and Schmidt-Ott [1998] tested a passenger car with a particulate trap and a truck powered by a diesel engine and equipped with an oxidation catalytic converter and a particulate trap. The type of filter medium, catalyst, and fuel used were not reported. Treatment of the passenger car exhaust gas in the particulate trap resulted in a large reduction of particulate mass and a significant increase in the number concentration of ultrafine particles. Small particles were found to be droplets of HC.

Carder [1999] tested a Rohmac-DCL system consisting of a monolith oxidation catalyst in line with a ceramic monolith particulate trap. The system was installed on an Isuzu C240 engine. The filtration system was designed for underground coal mine applications with the catalyst formulation selected to enhance regeneration. The formulation was not optimized for control of sulfate production. Diesel fuel sulfur content was 400 ppm (by weight). The weighted ISO 8178, eight-mode average DPM reduction for the system with the particulate trap placed downstream of the oxidation catalyst was reported to be 67.7%. The DPM reduction for the particulate trap with an oxidation catalyst installed upstream of it ranged from 40% to 99%, with an average of 78%. Therefore, the position of the trap and the DOC was found to have little effect on DPM emissions. The particulate trap with a downstream oxidation catalyst system also reduced HC and CO, on average, by 79% and 95%, respectively. The system with the catalyst upstream of the trap reduced HC and CO, on average, by 87% and 94%, respectively.

The Rohmac-DCL system was also tested on a Lister-Petter LPU-2 engine using the same fuel as in the work described above [Carder 1999]. The reductions in PM emissions of the trap-catalyst system were reported as 80% over modes 6 and 7. The average reductions of HC, CO, and $NO_x$ were found to be 97%, 90%, and 28%, respectively. MECA [1999a] found that when a DPF was used with an upstream NO-to-$NO_2$ catalyst and low-sulfur fuel (54 ppm), reductions in PM, HC, and CO emissions were 87%, 95%, and 93%, respectively.

Hansen et al. [2001] tested an in-use bus engine retrofitted with CRT system over 13 stationary modes (13-mode test) using 45 to 49 ppm of sulfur fuel. They found that sulfate, formed from sulfur in the fuel, is stored in the catalytic washcoat in the CRT filter's precatalyst at temperatures <380 °C and released again at temperatures >380 °C. The overall PM reduction efficiency of the tested CRT system measured over 13 modes was 55%. The system had a PM reduction efficiency >80% at the exhaust temperatures <375 °C. Hansen et al. found that sulfate, formed from sulfur in the fuel, is stored in the washcoat in the CRT filter precatalyst at temperatures <380 °C and released again at temperatures >380 °C. The analysis showed that sulfate and water constitute almost the entire particulate mass when the filter was used. In popular terms, the CRT filter just replaces carbon particles with sulfate particles. Hansen et al. concluded that reducing the sulfur content in the fuel cannot completely solve sulfate storage, but can prolong the time needed for the washcoat to become saturated with sulfur. They also concluded that even if ULSF is used, sulfur from the lubrication oil is sufficient to saturate the precatalyst. This is because 1% sulfur in the lubrication oil, which is not an unusual amount, corresponds to about 10-ppm sulfur in the fuel at normal rates of oil consumption.

### 2.4.4 Disposable Diesel Exhaust Filter (DDEF)

Disposable diesel exhaust filter (DDEF) systems are widely accepted by the underground coal mining industry [Ambs et al. 1994]. The DDEF system usually consists of a heat exchanger, filter element, filter housing, flame arrester, complete water jacketing to keep surface temperatures below MSHA requirements, exhaust temperature and exhaust back pressure monitor, and engine shutoff system. Heat exchangers are used to reduce the exhaust temperature to below 150 °C (dry exchangers) or 185 °C (water scrubber), one of several requirements for diesel equipment used in inby areas of underground coal mines [30 CFR 7 (1966)]. The reduced exhaust temperatures enable the use of disposable paper filters ($35 and $145). The filter elements presently used have a service life of about one to three shifts and cost from $35 to $145 depending on the system.

Since the maximum surface temperature of all surfaces of the permissible diesel engine and the diesel power package is limited by regulations to 150 °C (302 °F) [30 CFR 7 (1996)], exhaust system components of such a package are usually water-jacketed. Some efforts are now being made to replace bulky water-jacket systems with high-tech insulation materials.

Water scrubbers and dry-heat exchanger systems are the most commonly used systems for reducing exhaust temperature. The water scrubber has the dual purpose of cooling the exhaust and quenching flames and sparks at an acceptable engine back pressure. However, the water scrubber has high maintenance requirements due to the need for replenishing the water in the scrubber every few hours and the corrosion caused by the conversion of the $SO_2$ in the water box to sulfuric acid.

Dry scrubber systems use a heat exchanger to reduce exhaust temperatures at the filter face. This technology with an incorporated DOC has demonstrated the capability to reduce DPM by 97% compared to the engine output, when tested under the ISO 8178, eight-mode protocol in the laboratory [Paas 1999].

The major drawbacks of DDEF systems are high initial costs and large dimensions. However, these paper filter systems are currently the only filtration systems available for inby diesel equipment in underground coal mines.

### 2.4.4.1 Review of Published Results

Ambs et al. [1994] conducted a study on the performance of DDEFs. The study was done on a Jeffrey 4114 Ramcar powered by an MWM D916-6 engine and equipped with a DDEF system and waterbath exhaust conditioner (water scrubber). Field evaluation results showed that the DDEF reduced diesel exhaust aerosol concentrations in the mine ambient air from 70% to 90%. Ambs et al. found that the usable life of the filter ranged from 10 to 32 hr depending on factors such as mine altitude, engine type, and duty cycle.

Carder [1999] tested a Caterpillar 3306 engine retrofitted with a Dry System Technology (DST) dry scrubber system. The tests showed an 82% reduction in the averaged ISO 8178 eight-mode PM mass emission rate, but the reduction was rather low (8%) under rated speed/maximum load conditions. This indicated a low efficiency of the paper filter at high filter face temperatures, resulting from inability of the heat exchanger to cool the exhaust sufficiently.

## 3 Conclusions and Recommendations

Table 6 illustrates potential reductions in diesel emissions achievable by various control technologies that can be applied to curtail diesel emissions from underground equipment. Caution should be exercised when interpreting the emission numbers given in the examples. The numbers are based on published results from tests on similar technologies, but performed with different engines and under a variety of test conditions. The references for the data can be found in the text describing the particular technology applied. Where numerical data were not available, the performance evaluation is descriptive and not quantitative.

Extrapolation of these estimates to other applications (engines and/or duty cycles) is not exact. They should be considered as indicative of possible results and not necessarily as quantitatively accurate. The complexity of the problem requires a system to be designed and optimized for a particular application. Thus, the actual field performance of the combinations of the engines and control technologies may differ significantly from the estimates given in the table.

Selection of a particular technology is based on recognizing all of the peculiarities and interactions of the application. The choice is complex, based on engine operating conditions, reduction performance, and capital and operational costs. Further, present health concerns and worker exposure data indicate the need to focus on the reduction of PM emissions when choosing the appropriate technology. However, at the same time, the opportunity to reduce toxic gases should not be neglected. Fortunately, several combinations of technologies result in reductions of both PM and gases.

The use of a CDPF in combination with an FBC if needed for regeneration (DPF + FBC) seems to be the most effective aftertreatment technology for reductions of DPM and gaseous emissions from the diesel exhaust. The continuous regeneration of the DPF can be achieved for vehicles that operate at medium and high loads for at least 20% to 25% of the time. DPF + FBC combinations were reported to have efficiencies as high as 95% in the removal of DPM in both field and laboratory conditions. The use of ULSF is not mandatory in such systems, but can further reduce DPM mass emission and nanoparticle concentrations by eliminating sulfate formation.

DPFs were proven to be effective in removing ultrafine and nano-sized particles from the exhaust. When this combination is used with a new low-PM emitting engine, the resulting DPM emissions are further reduced in proportion to the ratio of the PM emissions (MSHA PI indices) of the two engines. Literature indicates that the potential of this latter combination is a reduction of DPM by 97%, resulting in *an additional 40% lower workplace DPM concentration* over the system that provides a 95% reduction. This combination (low-PI engine + DPF + FBC) provides good reductions of gaseous emissions as well.

If the reduction of gaseous emissions is the primary target, any combination of technologies using a DOCC is the most effective. The use of ULSF is recommended to prevent increases in the emission of sulfate particles. Substantial PM reductions (50%) can be obtained by the use of a water emulsion of the ULSF. Alternatively or in addition, replacement of the (older) engine with a low-emission engine can further reduce PM emissions up to 80%. The combination of a DOCC, water-fuel emulsions, and ULSF with or without low-emission engines is not well explored in the literature. The authors would welcome the opportunity to verify this promising option.

**Table 6.— Performance of the available control technologies**

NOTE: The effects on DPM emissions are reported on a mass basis, not a number basis. "Unknown" means that the authors have not found specific data for that diesel exhaust component, usually the SOL and SOF of the DPM. However, in all cases, the total effect on DPM is known.

| TECHNOLOGY | TOTAL DPM | | | | TOXIC GASES | | | COMMENTS |
| --- | --- | --- | --- | --- | --- | --- | --- | --- |
| | SOL | SOF | SULFATE | CO | NO$_2$ | HC | | |
| New low-emission engine | Unknown | Unknown | Not affected | Unknown | Unknown | Unknown | See table of MSHA-approved engines listed in table A-1. Extra initial capital cost may pay off in the long run in lower operating costs owing to fuel efficiency. |
| | Possibly >60% reduction based on MSHA certification testing, depending on engine | | | The MSHA ventilation rates indicate a reduction or increase in the gaseous emissions | | | |
| DOC + diesel No. 2 | No reductions | Up to 80% reduction | Possible significant increase, function of exhaust temperature | Reductions from 70% to 93% | Slight increase, strong function of exhaust temperature and catalyst formulation | Reductions from 60% to 75% | Diesel No. 2 has <500 ppm sulfur by weight, average 350 ppm. Sulfate generation has detrimental effects. Ultralow sulfur is recommended to prevent sulfate formation. DOCs are relatively inexpensive. |
| | Depends on SOF levels and fuel sulfur level one can expect either reduction or increase in total DPM. | | | | | | |
| DOC + ULSFs | No reductions | Reduction up to 80% | Insignificant presence | Same as DOC above | Same as DOC above | Same as DOC above | ULSFs are defined as those with <50 ppm sulfur per weight. Relatively low additional fuel costs make the technology viable. On-highway availability by mid-2006. |
| | Reduction up to 35% | | | | | | |

| TECHNOLOGY | TOTAL DPM | | | TOXIC GASES | | | COMMENTS |
|---|---|---|---|---|---|---|---|
| | SOL | SOF | SULFATE | CO | $NO_2$ | HC | |
| Biodiesel | Unknown | Reduction up to 48% | Not present | Reductions up to 40% | Significant increase. $NO_x$ up to 12% | Significant reductions | Biodiesel is oxygenated fuel that provides more complete combustion, generating less unburned HC and more $NO_x$ and $NO_2$. Significant reductions in PAHs observed. Available torque is reduced up to 7%. Higher fuel consumption up to 13%. Available, but $2.60/gal. |
| | Reduction up to 50% | | | | | | |
| Synthetic diesel (Fisher-Tropsch) | Unknown | Unknown | Not present | Reductions up to 33% | Unknown, reductions in $NO_x$ up to 27% | Reductions up to 49% | Significant reductions in PAHs. Small effects on available torque and fuel consumption. Industry is working on establishing economical production facilities. |
| | Reduction up to 21% | | | | | | |
| Water-fuel emulsions | Unknown | Unknown | Unknown | Increase | Unknown, reductions in $NO_x$ up to 30% | Increase | Simultaneously reduces PM and $NO_x$. Emulsion stability in storage, freezing, corrosion, and lubricity are concerns. Available and priced comparably to regular diesel fuel. |
| | Reduction up to 50% | | | | | | |

| TECHNOLOGY | TOTAL DPM | | | TOXIC GASES | | | COMMENTS |
| --- | --- | --- | --- | --- | --- | --- | --- |
| | SOL | SOF | SULFATE | CO | NO$_2$ | HC | |
| Fuel-borne catalyst (FBC) | Reduced | Increased | Not affected | Not affected | Not affected. reductions in NO$_x$ up to 15% | Not affected | Fuel economy might improve by 4.5%. Primary role of the FBC in mining is to lower DPF regeneration temperature. Because of significant emission of the nanoparticles, the FBC should not be used without DPF. |
| | Reductions up to 25% | | | | | | |
| Particulate filters (DPFs) | Unknown | Unknown | Not affected | Not significantly affected | Not significantly affected | Not significantly affected | High exhaust temperature required for autoregeneration can be reduced by catalysts. Slight increase in fuel consumption. |
| | Reductions from 80% to 95% | | | | | | |
| DOC + DPF | Up to 95% | Up to 80% | Slight increase | Reductions up to 93% | Slight to significant increase, strong function of exhaust temperature and catalyst formulation | Reductions up to 97% | The best reductions of DPM with DOC occurs with use of ULSF. |
| | Reductions up to 85% [Czerwinski et al. 1998] | | | | | | |

| TECHNOLOGY | TOTAL DPM | | | TOXIC GASES | | | COMMENTS |
|---|---|---|---|---|---|---|---|
| | SOL | SOF | SULFATE | CO | $NO_2$ | HC | |
| CDPF | Unknown | Unknown | Slight increase | Reductions up to 63% | Slight to significant increase | Reductions up to 94% | Depends on catalyst and temperatures. ignition temperatures around 450° C. Use of ULSF is recommended. |
| | Reductions up to 95% depending on sulfate | | | | | | |
| CDPF + FBC | Unknown | Unknown | Slight increase | Reductions up to 63% | Slight to significant increase | Reductions up to 94% | Ignition temperatures around 320° C. Fuel additive increases cost of fuel by up to 10 cents/gal. |
| | Reductions up to 95% depending on sulfate | | | | | | |
| CDPF + ULSF + FBC | Unknown | Unknown | Insignificant Presence | Reductions up to 63% | Slight to significant increase | Reductions up to 94% | Most effective retrofit technology. Ignition temperatures around 320° C. Fuel additive and lower sulfur increase cost of fuel by up to 20 cents/gal. |
| | Reductions up to 95% and no contribution from sulfates | | | | | | |
| Low-emission engine + DOC + No. 2 diesel fuel (350 ppm S) | Unknown | Significant reductions | Slight increase depending on DOC | Significant reductions | Slight to significant increase depending on DOC and engine | Significant reductions | Effective combination, yielding significant reductions in CO and HC (not $NO_2$) and fairly good reductions in PM. However DOC produces some sulfates that diminish the PM reductions possible. |
| | Reductions up to 87% compared to old technology engine baseline (up to 80% from engine, 35% reduction from DOC, but diminished by sulfate content) | | | | | | |
| Low-emission engine + DOC + ULSF | Unknown | Significant reductions | Insignificant Presence | Significant reductions | Slight to significant increase | Significant reductions | Very effective combination, yielding significant reductions in CO and HC (not $NO_2$) and fairly good reductions in PM. |
| | Reductions up to 87% compared to old technology engine baseline (up to 80% from engine, 35% from DOC + ULSF) | | | | | | |

| TECHNOLOGY | TOTAL DPM ||| TOXIC GASES ||| COMMENTS |
|---|---|---|---|---|---|---|---|
| | SOL | SOF | SULFATE | CO | $NO_2$ | HC | |
| Low-emission engine + DOC + ULSF + water emulsion | Unknown | Significant reductions | Insignificant presence | Significant reductions | Slight to significant increase | Significant reductions | Very effective combination yielding significant reductions in CO, HC, and $NO_x$ and good reductions in PM. |
| | Reductions up to 85% (estimate) |||||| |
| Low-emission engine + CDPF + No. 2 diesel fuel (350 ppm S) + FBC | Unknown | Significant reductions | Slight increase depending on DOC | Moderate reductions | Decrease reported, but increase possible | Moderate reductions | Excellent reduction in PM and good reduction in gaseous emissions. However, DOC produces some sulfates that diminish the PM reductions possible. |
| | Reductions up to 97%, but diminished by presence of sulfates |||||| |
| Low-emission engine + CDPF + ULSF + FBC | Unknown | Significant reductions | Insignificant presence | Moderate reductions | Decrease reported, but increase possible | Moderate reductions | The most effective combination results in excellent reduction in PM and good reduction in gaseous emissions. |
| | Reductions up to 97% |||||| |

## REFERENCES

66 Fed. Reg. 5526 and corrections 66 Fed. Reg 27864 [2001]. Mine Safety and Health Administration: 30 CFR 72, diesel particulate matter exposure of underground coal miners; final rule.

66 Fed. Reg. 5706 and corrections 66 Fed. Reg. 35518 [2001]. Mine Safety and Health Administration: 30 CFR 57, diesel particulate matter exposure of underground metal and nonmetal miners; final rule.

ACGIH [1995]. Notice of intended change: diesel particulate. Draft documentation of the proposed TLV for diesel exhaust. Cincinnati, OH: American Conference of Industrial Governmental Hygienists.

Ambs JL, Cantrell BK, Watts WF, Olson KS [1994]. Evaluation of a disposable diesel exhaust filter for permissible mining machines. Minneapolis, MN: U.S. Department of the Interior, Bureau of Mines, RI 9508.

Andersson J, Wedekind B [2001]. DETR/SMMT/CONCAWE particulate research programme: summary report. [http://www.ricardo.com/downloads/SummaryReport.pdf].

Bach E, Zikoridse G, Sandig R, Lemaire J, Mustel W, Naschke W, et al. [1998]. Combination of different regeneration methods for diesel particulate trap. Warrendale, PA: Society of Automotive Engineers, SAE paper 980541.

Bagley S, Gratz L [1998]. Evaluation of biodiesel fuel and oxidation catalysts in an underground metal mine, part 3—biological and chemical characterization. DEEP technical report. [http://www.deep.org].

Bagley ST, Gratz LD, Leddy DG, Johnson JH [1993]. Characterization of particle- and vapor-phase organic fraction emissions from a heavy-duty diesel engine equipped with a particle trap and regeneration control. Boston, MA: Health Effects Institute, Research Report No. 56.

Baranescu RA [1988]. Influence of fuel sulfur on diesel particulate emissions. Warrendale, PA: Society of Automotive Engineers, SAE paper 980541.

Baumgard KJ, Johnson JH [1996]. The effects of fuel and engine design on diesel exhaust particle size distributions. Warrendale, PA: Society of Automotive Engineers, SAE paper 960131.

Baz-Dresch JJ, Bickel KL, Watts WF [1993]. Evaluation of catalyzed diesel particulate filters in an underground metal mine. Minneapolis, MN: U.S. Department of the Interior, Bureau of Mines, RI 9478.

Bugarski AD [1999]. Characterization of particulate matter and hydrocarbon emissions from in-use heavy-duty diesel engine [Dissertation]. Morgantown, WV: West Virginia University, Department of Mechanical and Aerospace Engineering.

Burtsher H, Skillas G, Baltensperger U, Matter U [1999]. Particle formation due to fuel additives, 3. International ETH-Workshop on Nanoparticle Measurement (Zürich, Switzerland, August 9-10, 1999).

Cantrell BK, Williams KL, Watts WF, Jankowski RA [1993]. Mine aerosol measurement. In: Willeke K, Baron PA, eds. Aerosol measurement: principles, techniques, and applications. Van Nostrand, pp. 591-611.

Carder DK [1999]. Performance evaluation of exhaust aftertreatment devices used for emissions control on diesel engines employed in underground coal mines [Thesis]. Morgantown, WV: West Virginia University, Department of Mechanical and Aerospace Engineering.

CFR. Code of Federal Regulations. Washington, DC: U.S. Government Printing Office, Office of the Federal Register.

Corning, Inc. [2000]. Corning environmental technologies. [http://www.corning.com/environmentaltechnologies].

Cowley LT, Jeune AL, Lange WW [1993]. The effect of fuel composition including aromatics content on emissions from a range of heavy-duty diesel engines. In: Fourth International Symposium on the Performance Evaluation of Automotive Fuels and Lubricants (CEC, Birmingham, U.K.).

Czerwinski J, Mosimann T, Matter U, Kasper M [1998]. Investigation with the HJS-CRT system on the Liebherr D914T construction engine with detailed analysis of the particulate emissions. TTM. Unpublished (proprietary test report).

DECSE [1999]. Phase I interim data report No. 3: diesel fuel sulfur effects on particulate matter emissions. [http://www.ott.doe.gov/decse].

DECSE [2000]. Phase I interim data report No. 4: diesel particulate filters—final report. [http://www.ott.doe.gov/decse].

DECSE [2001]. Diesel emission control–sulfur effects (DECSE) final report. [http://www.ott.doe.gov/decse].

Den Ouden CJJ, Clark RH, Cowley LT, Stradling RJ, Lange WW, Maillerd C [1994]. Fuel quality effects on particulate matter emissions for light- and heavy-duty diesel engines. Warrendale, PA: Society of Automotive Engineers, SAE paper 942022.

DieselNet [1997a]. Ceramic catalyst substrates, DieselNet technology guide. [http://www.DieselNet.com/tg.html].

DieselNet [1997b]. Metallic catalyst substrates, DieselNet technology guide. [http://www.DieselNet.com/tg.html].

DieselNet [1998a]. Cellular monolith substrates, DieselNet technology guide. [http://www.DieselNet.com/tg.html].

DieselNet [1998b]. Clean diesel engine—emission control technologies, DieselNet technology guide. [http://www.DieselNet.com/tg.html].

DieselNet [1998c]. Diesel catalysts: deactivation of diesel catalyst, DieselNet technology guide. [http://www.DieselNet.com/tg.html].

DieselNet [1998d]. Fuel properties and emissions, diesel fuels, DieselNet technology guide. [http://www.DieselNet.com/tg.html].

DieselNet [1999a]. Diesel oxidation catalyst. DieselNet technology guide. [http://www.DieselNet.com/tg.html].

DieselNet [1999b]. Exposure to diesel exhaust. DieselNet technology guide. [http://www.DieselNet.com/tg.html].

DieselNet [2001]. Emission standards. [http://www.DieselNet.com/standards.html].

Duffy JS, Samarchi S [1997]. The impact of platinum in diesel exhaust on human health. Final report prepared for Clean Diesel Technologies, Inc., Stamford, CT.

Durbin TD, Collins JR, Norbeck JM, Smith MR [2000]. Effects of biodiesel, biodiesel blends, and a synthetic diesel on emissions from light heavy-duty diesel vehicles. Env Sci Tech *34*(3):349-355.

EMA [1999]. EMA study estimates cost increase to produce low sulfur diesel fuel. [http://www.DieselNet.com/news/9911ema.html].

Engelhard Corp. [1999]. Emission control products. [http://www.DieselNet.com/engelhard/products.html].

EPA [2001]. Biography of biodiesel studies. [http://www.epa.gov/otaq/models/analysis/biodsl/bibliog.pdf].

Forbush S [2001]. Diesel emissions reduction program. Unpublished paper presented at the Mining Diesel Emissions Conference, Markham, Ontario, Canada, November 7.

Gangal MK, Dainty ED [1993]. Ambient measurement of diesel particulate matter and respirable combustible dust in Canadian mines. In: Bhaskar R, ed. Proceedings of the Sixth U.S. Mine Ventilation Symposium. Littleton, CO: Society for Mining, Metallurgy, and Exploration, Inc., pp. 83-89.

Gangal M, Ebersole J, Vallieres J, Dainty D [1990]. Laboratory study of current (1990/91) soot/RCD sampling methodology for the mine environment. Ottawa, Ontario, Canada: Canada Centre for Mineral and Energy Technology (CANMET), Mining Research Laboratory.

Haney RA, Saseen GP, Waytulonis RW [1997]. An overview of diesel particulate exposures and control technology in the U.S. mining industry. Appl Occup Env Hyg $12$:1013-1018.

Hansen KF, Jensen MG, Ezerman N [2001]. Measurement of the "true" efficiency of a CRT filter. Report to Danish Road Safety and Transport Agency. [http://www.fstyr.dk/udvikling/Odense/Rap_partikel/rap_eng.pdf].

HEI [2001]. Evaluation of human health risk from cerium added to diesel fuel. Boston, MA: Health Effects Institute, Communication 9. [http://www.healtheffects.org/Pubs/Cerium.pdf].

Heywood JB [1988]. Internal combustion engine fundamentals. New York, NY: McGraw-Hill, Inc.

Houben H, Miebach R, Sauerteig JE [1994]. The optimized Deutz service diesel particulate filter system DPFS II. Warrendale, PA: Society of Automotive Engineers, SAE paper 942264.

Howard JB, Kausch WJ Jr. [1980]. Soot control by fuel additives. Prog Energy Combust Sci $6$:263.

Howell S, Weber AJ [1997]. Biodiesel use in underground metal and nonmetal mines. DieselNet technical reports. [http://www.DieselNet.com/papers/9705howell.html].

Jelles SJ, Makkee M, Moulijn JA, Acres GJK, Peter-Hoblyn JD [1999]. Diesel particulate control. Application of an activated particulate trap in combination with fuel additives at an ultra low dose rate. Warrendale, PA: Society of Automotive Engineers, SAE paper 1999-010-0113.

Kahlert B [1999]. First use of additive-regenerated diesel particulate filters in Germany mines. In: Proceedings of the Mining Diesel Emissions Conference (Markham, Ontario, Canada).

Kauffeldt TH, Schmidt-Ott A [1998]. Investigation of the influence of exhaust aftertreatment on the particulate phase in diesel exhaust gases. J Aerosol Sci $29$(Suppl1):S625-S626.

Kittelson DB [1998]. Engine and nanoparticles: a review. J Aerosol Sci $29$(5/6):575-588.

Klein H, Lox E, Kreuzer T, Kawanami M, Ried T, Bächmann K [1998]. Diesel particulate emissions of passenger cars: new insights into structural changes during the process of exhaust aftertreatment using diesel oxidation catalysts. Warrendale, PA: Society of Automotive Engineers, SAE paper 980196.

Langer AD, Daly TD [1999]. Low-emission water blend diesel fuel. In: Proceedings of the Mining Diesel Emissions Conference (Markham, Ontario, Canada).

Larsen CA, Levendis YA, Shimato K [1999]. Filtration assessment and thermal effects of aerodynamic regeneration in silicon carbide and cordierite particulate filters. Warrendale, PA: Society of Automotive Engineers, SAE paper 1999-01-0466.

Lepperhoff G, Lüders H, Barthe P, Lemaire J [1995]. Quasi-continuous particle trap regeneration by cerium additives. Warrendale, PA: Society of Automotive Engineers, SAE paper 950369.

Lubrizol [1999]. Lubrizol to develop diesel fuel-water blends. DieselNet News. [http://www.DieselNet.com/news/9908lubrizol2.html].

Majewski A [1999]. Current trends in diesel emission regulations. In: Proceedings of the Mining Diesel Emissions Conference (Markham, Ontario, Canada).

Maskery D [1978]. Respirable combustible dust. INCO method No. 1H011A.

Mauderly J, Schlesinger R, Neas L [1995]. Measurement needs related to health effects. Boston, MA: Health Effects Institute.

Mayer A [1997]. VERT: curtailing emissions of diesel engines in tunnel sites. Technical report to VERT. TTM.

Mayer A, Egli H, Burtscher H, Czerwinski W, Gehrig D [1995]. Particle Size distribution downstream traps of different design. Warrendale, PA: Society of Automotive Engineers, SAE paper 950373.

Mayer A, Matter U, Czerwinski J, Heeb N [1999]. Effectiveness of particulate traps on construction site engines: VERT final measurements. DieselNet technical report. [http://www.DieselNet.com/papers/9903mayer/index.html].

McClure BT, Baumgard KJ, Watts WF [1988]. Effectiveness of catalytic converters on diesel engines used in underground mining. Minneapolis, MN: U.S. Department of the Interior, Bureau of Mines, IC 9197.

McCormick RL, Graboski MS, Alleman TL, Herring AM [2001]. Impact of biodiesel source material and chemical structure on emissions of criteria pollutants from a heavy-duty engine. Env Sci & Tech *35*(9).

McGinn S [1999]. Maintenance guidelines and best practices for diesel engines. [http://www.deep.org/reports/mtce_guidelines.pdf].

McGinn S [2000]. The relationship between diesel engine maintenance and exhaust emissions—final report. http://www.deep.org/reports/mtce_report.pdf].

McGinn S [2001a]. Brunswick mine particulate trap project: performance evaluation. In: Proceedings of the Mining Diesel Emissions Conference (Markham, Ontario, Canada).

McGinn S [2001b]. Post session public discussion on the DEEP trap evaluation program at Brunswick Mining and Smelting operations. In: Proceedings of the Mining Diesel Emissions Conference, Markham, Ontario, Canada).

McGinn S, Weeks S, Gaultier P [2000]. Diesel engine maintenance audit plan. [http://www.deep.org/reports/mtce_audit.pdf].

McMillian MH, Gautam M [1998]. Consideration for Fisher-Tropsch derived liquid fuels as a fuel injection emission control parameter. Warrendale, PA: Society of Automotive Engineers, SAE paper 982489.

MECA [1999a]. Demonstration of advanced emission control technologies enabling diesel-powered heavy-duty engines to achieve low emission levels. Final report. [http://www.DieselNet.com/news/9907meca.html].

MECA [1999b]. MECA calls for 30-ppm sulfur fuel. DieselNet News. [http://www.DieselNet.com/news/9903meca.html].

MSHA [1997]. Practical ways to reduce exposure to diesel exhaust in mining—a toolbox. [http://www.msha.gov/S&HINFO/TOOLBOX/TBCOVER.HTM].

MSHA [2001a]. Diesel particulate matter (DPM) control technologies with percent removal efficiency. [http://www.msha.gov/01-995/Coal/DPM-FilterEfflist.pdf].

MSHA [2001b]. MSHA-approved diesel engines, powerpackages, and equipment. [http://www.msha.gov/S&HINFO/DESLREG/APPENG.HTM].

National Biodiesel Board [2001]. [http://www.biodiesel.org].

Nault G [2001]. Diesel particulate filter study at INCO's Stobie mine. Unpublished paper presented at the Mining Diesel Emissions Conference, Markham, Ontario, Canada.

NIOSH [1999]. Elemental carbon (diesel particulate): method 5040, issue 3 (interim). In: NIOSH manual of analytical methods. 4th rev. ed. [http://www.cdc.gov/niosh/nmam/pdfs/5040f3.pdf].

Norton P, Vertin K, Clark NN, Lyons DW, Gautam M, Goguen S, et al. [1999]. Emissions from buses with DDC 6V92 engines using synthetic diesel fuel. Warrendale, PA: Society of Automotive Engineers, SAE paper 1999-01-1512.

Paas N [1999]. Dry system technology management system. In: Proceedings of the Mining Diesel Emissions Conference (Markham, Ontario, Canada).

Pataky GM, Baumgard KJ, Gratz LD, Bagley ST, Leddy DG, Johnson JH [1994]. Effects of an oxidation catalytic converter on regulated and unregulated diesel emissions. Warrendale, PA: Society of Automotive Engineers, SAE paper 940243.

Petkewich R [2001]. Biodiesel at public pumps. Env Sci & Tech *Aug 1*:321A.

Popuri S, Gautam M, Rankin B, Seehra M [1999]. Development of a microwave-assisted regeneration system for a ceramic diesel particulate system. Warrendale, PA: Society of Automotive Engineers, SAE 1999-01 3565.

Postnikoff JA [1999]. Diesel particulate matter minimization at Agrium. In: Proceedings of the Mining Diesel Emissions Conference (Markham, Ontario, Canada).

Schaberg PW, Myburgh IS, Botha, JJ, Roets PN, Viljoen CL, Dancuart LP, et al. [1997]. Diesel exhaust emissions using Sasol slurry phase distillate process fuels. Warrendale, PA: Society of Automotive Engineers, SAE paper 972898.

Schlesinger J [1995]. Toxicological evidence for health effects from inhaled particulate pollution: does it support the human experience? Inhal Toxicol *7*:99-109.

Schnakenberg GH Jr. [1999a]. Estimates of achievable workplace DPM levels. In: Proceedings of the Mining Diesel Emissions Conference (Markham, Ontario, Canada).

Schnakenberg GH Jr. [1999b]. Information obtained at DEEP trap bidders meeting, Markham, Ontario, Canada, November 2, 1999. Unpublished.

Sharp CA [1998]. Exhaust emissions and performance of diesel engines with biodiesel fuel. [http://www.biodiesel.org].

Spears MW [1997]. An emissions-assisted maintenance procedure for diesel-powered equipment. Minneapolis, MN: University of Minnesota, Center for Diesel Research. NIOSH contract No. USDI/1432 CO369004. [http://www.cdc.gov/niosh/mining/eamp/eamp.html].

Unikat AB [1999]. Engine Control Systems, Ltd. Canada, data sheet 7500. [http://www.unikat.se/fiter/html].

Watts WF, Cantrell BK, Bickel KL, Olson KS, Rubow KL, Baz-Dresch JJ, et al. [1995]. In-mine evaluation of catalyzed diesel particulate filters at two underground metal mines. Minneapolis, MN: U.S. Department of the Interior, Bureau of Mines, RI 9571.

Watts WF, Spears M, Johnson J [1998]. Evaluation of biodiesel fuel and oxidation catalysts in an underground metal mine. DEEP technical report. [http://www.deep.org].

Waytulonis R [1987]. An overview of the effects of diesel engine maintenance on emissions and performance. In: Diesels in Underground Mines; Proceedings: Bureau of Mines Technology Transfer Seminar, Louisville, KY, April 21, 1987, and Denver, CO, April 23, 1987. Minneapolis, MN: U.S. Department of the Interior, Bureau of Mines, IC 9141.

Xiaobin L, Chipior WL, Gulder OL [1996]. Effects of fuel properties on exhaust emissions of a single-cylinder DI diesel engine. Warrendale, PA: Society of Automotive Engineers, SAE paper 962116.

ZH 1/120.44 [1995]. Verfahren zur Bestimmung fon Kohlenstoff im Feinstaub—anwendbar für partikelförmige Dieselmotor-Emissionen in Arbeitsbereichen (in German). Köhn, Germany: Carl Heymanns Verlag KG.

## APPENDIX A.—LOW-PI MSHA-APPROVED ENGINES

Table A-1 lists the MSHA-approved engines (as of January 2, 2001) for nonpermissible areas. Shaded entries indicate an engine with a significantly lower PI than those of comparable horsepower. Although in some cases an engine choice may not be physically possible, an engine with the lowest PI that meets the physical size and approximate horsepower for the application should be chosen. For example, one should consider the 28.2-hp Deutz F2L1011 (PI=1,000) over either the Lister-Petter, LPU3 MKI (PI=7,000), LPU3 MKII (PI=4,500), or Deutz F2L1011F (PI=3,500). Similarly, the Deutz F4L1011 56.3 hp (PI=2,000) could replace the Isuzu C240 (PI=5,500). In the latter example, the tailpipe DPM is reduced by 64%.

### Table A-1.—List of MSHA-approved engines (as of December 2001)

NOTE: Shaded entries indicate an engine with a significantly lower PI than those of comparable horsepower.

| Approval No. | Engine model | hp @ rpm | MSHA nameplate ventilation rate, cfm | MSHA particulate index, cfm |
|---|---|---|---|---|
| 7E-B070-0 | Farymann Diesel, 43F | 14 @ 3,000 | 1,000 | 4,000 |
| 7E-B042-0 | Lister-Petter, LPU2 MKI[1] | 17.5 @ 3,000 | 1,000 | 5,000 |
| 7E-B053-0 | Kubota, Model V1200 | 25.8 @ 3,000 | 1,000 | 1,500 |
| 7E-B041-0 | Lister-Petter, LPU3 MKI[1] | 26.3 @ 3,000 | 1,500 | 7,000 |
| 7E-B062-0 | Deutz, F2L1011[1] | 28.2 @ 3,000 | 1,500 | 1,000 |
| 7E-B044-0 | Lister-Petter, LPU3 MKII | 29 @ 3,000 | 1,500 | 4,500 |
| 7E-B015-0 | Deutz F2L 1011F | 30 @ 3,000 | 2,000 | 3,500 |
| 7E-B091-0 | Deutz F2L1011F | 30 @ 3,000 | 2,000 | 1,500 |
| 7E-B074-0 | Isuzu 3LD1MA | 33.3 @ 3,000 | 2,000 | 3,500 |
| 7E-B040-0 | Lister-Petter, LPU4 MKI[1] | 35 @ 3,000 | 2,000 | 9,500 |
| 7E-B043-0 | Lister-Petter, LPU4 MKII | 38.6 @ 3,000 | 2,000 | 6,000 |
| 7E-B026-0 | Deutz, F3L912W (2.8L)[1] | 40 @ 2,300 | 2,500 | 2,500 |
| 7E-B076-0 | Isuzu 4LC1MA | 41 @ 3,000 | 2,000 | 4,500 |
| 7E-B061-0 | Deutz, F3L1011[1] | 41.6 @ 3,000 | 2,500 | 1,500 |
| 7E-B033-0 | Perkins, 104-19 | 42.5 @ 2,800 | 2,000 | 7,000 |
| 7E-B014-0 | Deutz F3L1011F | 44 @ 3,000 | 2,500 | 5,000 |
| 7E-B090-0 | Deutz F3L1011F | 44 @ 3,000 | 3,000 | 2,500 |
| 7E-B054-0 | Deutz, Model F3M1011F | 46 @ 2,800 | 3,000 | 3,500 |
| 7E-B031-0 | Deutz, F3L912W (3.1L)[2] | 47 @ 2,500 | 2,500 | 2,500 |
| 7E-B071-0 | Kubota Engine Corp., Model V2203-E | 48.4 @ 2,800 | 2,500 | 2,000 |
| 7E-B079-0 | New Holland North America, 201, 3.3L NA | 51 @ 2,200 | 4,500 | 8,500 |

| Approval No. | Engine model | hp @ rpm | MSHA nameplate ventilation rate, cfm | MSHA particulate index, cfm |
|---|---|---|---|---|
| 7E-B086-0 | Isuzu C240MA | 52 @ 3,000 | 3,000 | 2,500 |
| 7E-B025-0 | Deutz, F4L912W (3.8L)[1] | 54 @ 2,300 | 3,000 | 3,500 |
| 7E-B075-0 | Isuzu 4LE1MA | 54 @ 3,000 | 2,500 | 6,500 |
| 7E-B085-0 | Isuzu, C240MA | 56 @ 3,000 | 2,500 | 5,500 |
| 7E-B038-0 | Isuzu, C240MA (QD60)[1] | 56 @ 3,000 | 2,500 | 5,500 |
| 7E-B060-0 | Deutz, F4L1011[1] | 56.3 @ 3,000 | 3,000 | 2,000 |
| 7E-B027-0 | Perkins, 704-26 | 58 @ 2,600 | 2,000 | 8,000 |
| 7E-B089-0 | Deutz F4L1011F | 59 @ 3,000 | 4,000 | 3,000 |
| 7E-B013-0 | Deutz F4L1011F | 59 @ 3,000 | 3,500 | 6,500 |
| 7E-B089-0 | Deutz F4L1011F | 59 @ 3,000 | 4,000 | 3,000 |
| 7E-B077-0 | Deutz BF3M1011F[2] | 61 @ 2,800 | 5,500 | 5,500 |
| 7E-B055-0 | Deutz, Model F4M1011F | 61 @ 2,800 | 3,500 | 4,500 |
| 7E-B029-0 | Deutz, F4L912W (4.1L)[2] | 62 @ 2,500 | 3,000 | 3,500 |
| 7E-B024-0 | Deutz, F5L912W (4.7L)[1] | 67 @ 2,300 | 4,000 | 4,500 |
| 7E-B019-0 | Deutz BF4L 1011F | 74 @ 2,800 | 5,500 | 4,500 |
| 7E-B030-0 | Deutz, F5L912W (5.1L)[2] | 76 @ 2,500 | 4,000 | 4,000 |
| 7E-B006-0 | Isuzu QD 100-301[1,2] | 79 @ 2,800 | 5,000 | 8,500 |
| 7E-B023-0 | Deutz, F6L912W (5.6L)[1] | 80 @ 2,300 | 4,500 | 5,000 |
| 7E-B056-0 | Deutz, BF4M1011F[2] | 82 @ 2,800 | 5,500 | 5,500 |
| 7E-B072-0 | Detroit Diesel Series, Model DDC D704LTE | 84 @ 2,600 | 6,500 | 7,000 |
| 7E-B028-0 | Deutz, F6L912W (6.1L)[2] | 93 @ 2,500 | 4,500 | 5,000 |
| 7E-B001-0 | Deutz-MWM 916-6[1] | 94 @ 2,300 | 4,000 | 11,500 |
| 7E-B004-0 | Caterpillar 3304 PCNA[1] | 100 @ 2,200 | 5,000 | 15,000 |
| 7E-B022-0 | Perkins, 1004-40T[2] | 108 @ 2,400 | 9,000 | 9,000 |
| 7E-B064-0 | Caterpillar, 3054 DIT[2] | 108 @ 2,400 | 9,000 | 9,000 |
| 7E-B011-0 | Deutz BF4M1012C[2] | 110 @ 2,500 | 6,500 | 4,000 |
| 7E-B045-0 | Isuzu, 4BGIT-MA | 111 @ 2,400 | 7,000 | 13,000 |
| 7E-B011-0 | Deutz BF4M1012EC[2] | 113 @ 2,500 | 6,500 | 4,000 |
| 7E-B084-0 | Cummins 4BTA3.9-C | 116 @ 2,500 | 6,500 | 7,500 |
| 7E-B088-0 | Isuzu CBG1-MA1 | 116 @ 2,500 | 4,500 | 11,500 |
| 7E-B020-0 | Perkins, 1004-40TW | 122 @ 2,300 | 10,000 | 7,500 |
| 7E-B065-0 | Caterpillar, 3054 DIT | 122 @ 2,300 | 10,000 | 7,500 |
| 7E-B073-0 | Detroit Diesel Series, Model DDC D706LTE | 123 @ 2,600 | 7,500 | 11,000 |
| 7E-B059-0 | Deutz, BF4M1013E | 125 @ 2,300 | 11,500 | 4,500 |
| 7E-B059-0 | Deutz, BF4M1013 | 127 @ 2,300 | 11,500 | 4,500 |

| Approval No. | Engine model | hp @ rpm | MSHA nameplate ventilation rate, cfm | MSHA particulate index, cfm |
|---|---|---|---|---|
| 7E-B046-0 | Isuzu, 6BGI-MA | 129 @ 2,500 | 6,000 | 16,000 |
| 7E-B039-0 | Isuzu, 6BD1MA (QD145)[1] | 135 @ 2,800 | 9,000 | 12,000 |
| 7E-B034-0 | Deutz, F6L413FW | 137 @ 2,300 | 8,000 | 7,000 |
| 7E-B003-0 | Caterpillar 3306 PCNA[1] | 150 @ 2,200 | 7,500 | 23,000 |
| 7E-B021-0 | Perkins, 1006-60T | 152 @ 2,200 | 13,000 | 12,000 |
| 7E-B066-0 | Caterpillar, 3056 DIT | 152 @ 2,200 | 13,000 | 12,000 |
| 7E-B008-0 | Deutz BF4M1013C[2] | 158 @ 2,300 | 8,500 | 7,500 |
| 7E-B005-0 | General Motors L57, 6.5L Hummer | 160 @ 3,400 | 7,500 | 9,500 |
| 7E-B052-1 | Cummins Model B5.9,[2] without DOC | 160 @ 2,500 | 11,500 | 5,000 |
| 7E-B052-3 | Cummins Model B5.9,[2] with DOC | 160 @ 2,500 | 12,000 | 3,500 |
| 7E-B052-2 | Cummins Model B5.9,[2] without DOC | 175 @ 2,500 | 11,500 | 5,000 |
| 7E-B052-4 | Cummins Model B5.9,[2] with DOC | 175 @ 2,500 | 12,000 | 3,500 |
| 7E-B052-5 | Cummins Model B5.9,[2] with DOC | 180 @ 2,500 | 12,000 | 3,500 |
| 7E-B035-0 | Deutz, F8L413FW | 182 @ 2,300 | 10,500 | 9,500 |
| 7E-B067-0 | Navistar, A185, Model years 1988-1994 | 185 @ 3,300 | 8,500 | 15,000 |
| 7E-B058-0 | Deutz, BF6M1013E | 189 @ 2,300 | 17,500 | 5,500 |
| 7E-B058-0 | Deutz, BF6M1013 | 194 @ 2,300 | 17,500 | 5,500 |
| 7E-B016-0 | General Motors, L65, 6.5L, Turbo-Hummer | 195 @ 3,400 | 9,500 | 24,000 |
| 7E-B052-0 | Cummins Model B5.9[2] | 215 @ 2,700 | 14,500 | 4,000 |
| 7E-B052-6 | Cummins, Model B5.9,[2] with DOC | 215 @ 2,000 | 14,500 | 4,000 |
| 7E-B068-0 | Navistar, A215 and A225, Model years 1994.5-1997 | 215 @ 3,000 | 18,000 | 11,000 |
| 7E-B036-0 | Deutz, F10L413FW | 228 @ 2,300 | 13,500 | 12,000 |
| 7E-B057-0 | Deutz, BF6M1013EC | 228 @ 2,300 | 16,000 | 8,500 |
| 7E-B050-0 | Detroit Diesel Series, 40 Model N063DH32[2] | 230 @ 2,200 | 12,500 | 4,500 |
| 7E-B057-0 | Deutz, BF6M1013C | 233 @ 2,300 | 16,000 | 8,500 |
| 7E-B051-0 | Cummins, Model ISB-235 | 235 @ 2,700 | 10,000 | 6,000 |
| 7E-B069-0 | Navistar, A250, B235, and B250 | 250 @ 2,600 | 15,000 | 6,000 |
| 7E-B080-0 | Detroit Diesel Series 40, N063-DH32 I-308[2] | 250 @ 2,200 | 16,500 | 3,000 |
| 7E-B007-0 | Deutz BF6M1013C[2] | 261 @ 2,300 | 12,000 | 19,000 |
| 7E-B010-1 | Caterpillar, 3306 DITA[1,2] | 270 @ 2,200 | 15,000 | 6,000 |
| 7E-B037-0 | Deutz, F12L413FW | 274 @ 2,300 | 16,000 | 14,000 |
| 7E-B083-0 | Daimler Chrysler | 275 @ 2,200 | 11,000 | 7,000 |

| Approval No. | Engine model | hp @ rpm | MSHA nameplate ventilation rate, cfm | MSHA particulate index, cfm |
|---|---|---|---|---|
| 7E-B078-0 | Deutz BF6M1013FC-MVS | 282 @ 2,300 | 17,000 | 4,000 |
| 7E-B017-0 | Caterpillar, 3306 ATAAC[1] | 300 @ 2,200 | 11,500 | 12,000 |
| 7E-B047-0 | Detroit Diesel Series, 50 DDFC[1,2] | 315 @ 2,100 | 16,000 | 5,000 |
| 7E-B092-0 | Detroit Diesel Series, 50 Model 6043TK32[2] | 315 @ 2,100 | 21,000 | 7,000 |
| 7E-B048-0 | Detroit Diesel Series, 60 (11.1L)DDEC[1,2] | 325 @ 2,100 | 18,000 | 5,500 |
| 7E-B012-0 | Caterpillar 3176 ATAAC[2] | 335 @ 2,100 | 15,000 | 8,000 |
| 7E-B002-0 | Deutz BF6M1015C[2] | 402 @ 2,100 | 18,500 | 17,500 |
| 7E-B049-0 | Detroit Diesel Series, 60 (12.7L)DDEC[1,2] | 475 @ 2,100 | 28,000 | 8,500 |
| 7E-B018-0 | Caterpillar, 3406E ATAAC[1,2] | 500 @ 2,100 | 24,000 | 12,500 |
| 7E-B082-0 | Caterpillar, 3408E DITA | 510 @ 2,000 | 33,000 | 17,000 |
| 7E-B009-0 | Deutz BF8M1015C[2] | 536 @ 2,100 | 24,000 | 18,000 |
| 7E-B087-0 | Detroit Diesel Series 60 14L DDEC IV Model 6063HK32[2] | 575 @ 3,000 | 28,000 | 12,000 |
| 7E-B081-0 | Cummins QSK19-C | 650 @ 2,100 | 45,000 | 33,000 |
| 7E-B032-0 | Detroit Diesel, 8V-2,000TA DDEC | 650 @ 2,100 | 45,000 | 10,000 |

[1]Engine was previously approved under Part 32.
[2]Lower horsepower ratings have been approved.

Because the PI increases with engine horsepower (just as the ventilation rate does), it is easier to compare engines by examining their rate of particulate production per engine horsepower—that is, by comparing the number obtained by dividing the PI by the rated horsepower of each engine chosen. For example, the 175-hp Cummins engine (7E-B052-4) has a PI/hp of 20 cfm/hp (3,500 cfm/175 hp). Using this as a benchmark, of the remaining 21 MSHA-approved engines rated between 200 and 650 hp, only 7 have a PI/hp <20 cfm/hp; 5 have a PI/hp between 21 and 30; 3 between 31 and 40; 2 between 41 and 50; and only 1 at 72.8 cfm/hp. The range of PI/hp for the engines listed above is 19.6 to 285.7 cfm/hp.

www.ingramcontent.com/pod-product-compliance
Lightning Source LLC
Chambersburg PA
CBHW081901170526
45167CB00007B/3098